Enduring Pastoral

Architecture Technology Culture 4

Editorial Address:
Prof. Dr. Klaus Benesch
Department of English/American Studies
University of Munich
Schellingstrasse 3/VG
80799 Munich

Enduring Pastoral

Recycling the Middle Landscape Ideal in the Tennessee Valley

Torben Huus Larsen

Rodopi

Amsterdam - New York, NY 2010

Illustration cover: Chattanooga Choo Choo, 2006.
Photography by Torben Huus Larsen.

Cover design: Pier Post

The paper on which this book is printed meets the requirements of "ISO 9706: 1994, Information and documentation - Paper for documents - Requirements for permanence."

ISBN: 978-90-420-3057-2
E-Book ISBN: 978-90-420-3058-9

Printed in The Netherlands

For my parents, Bente & Steen Larsen, and my sister, Katrine.

Acknowledgments

I owe a debt of gratitude to many people in relation to this project. First and foremost I must thank Professor David E. Nye, whose knowledge and guidance has been nothing short of invaluable over the years and who has helped me to constantly rethink my approach to the subject at hand. I am further indebted to the members of my Ph.D. dissertation committee, Associate Professor Jørn Brøndal, Senior Lecturer Steven Hartman, and Professor Klaus Benesch, for offering important criticism and advice. I would also like to show my appreciation to my colleagues at the University of Southern Denmark who have proven an inspiration and a helpful sounding board, as well as to the Department of History and Civilization at SDU for giving me the opportunity to write this book.

During my travels in the Tennessee Valley I have visited a number of libraries and archives, and I have experienced nothing but understanding and interest in my research. Twice I have visited the Southeast Region of the National Archives just outside of Atlanta in Morrow, Georgia, and on both occasions I had the good fortune of working with Ms. Arlene Royer, whose patience and knowledge of the files relating to the Tennessee Valley Authority was priceless. Without the aid from Ms. Royer and the rest of the staff, I would undoubtedly still be at the archives buried in boxes and documents.

In Columbus, Mississippi, Ms. Agnes Zaiontz and Mr. Donald Waldon from the Tennessee-Tombigbee Waterway Development Authority took time out of their busy schedules not only to talk to me about the Tennessee-Tombigbee Waterway, its history, and present role in the local community, but to introduce me to their wonderful town. Their commitment and candor was inspiring and I owe them thanks for their time and effort.

I would also like to thank the people working at the Mitchell Memorial Library at Mississippi State University in Starkville, Mississippi, the James D. Hoskins Library at the University of Tennessee in Knoxville, the Columbus-Lowndes Public Library in Columbus, Mississippi, the Chattanooga-Hamilton County Bicentennial Library in Chattanooga, Tennessee, the staff at the Pack Memorial Library in Asheville, and the library at the University of Southern Denmark.

Contents

Introduction

Located in Pigeon Forge, Tennessee, in the foothills of the Great Smoky Mountains, Dollywood is one of the Tennessee Valley's largest and best known tourist attractions. Part amusement park and part showcase for regional history and culture, the theme park, which is partly owned by singer Dolly Parton, sells a vision of down-home country idyll. By blending antique technologies such as steam-driven locomotives and water-powered grist mills with fountains, Smoky Mountain scenery, and colorful flowerbeds, the park is meant to entertain as well as to purvey a sense of place and history. The many stage shows focus on local styles of music, including country, bluegrass, and mountain gospel music. A variety of shops display working demonstrations of traditional arts and crafts, and the park's restaurants serve traditional regional foods. Adding a sense of piety to the scene is an old wooden church, suggesting to tourists a bygone era when spirituality was a more integrated part of life. At the end of the day, if visitors leave the park feeling nostalgic, it is because Dollywood's dominant narrative emphasizes the rusticity of country living and a way of life that seemingly has been lost as a result of modernity.

Dollywood, of course, is one big contradiction. The images of rusticity are scattered amongst state-of-the-art rollercoasters, souvenir shops, and the thousands of tourists who have chosen to spend a day in the park. Within such a hybridized and commercial context one should think it impossible, for example, for a replica of Dolly Parton's childhood home to imbue the park with a sense of authenticity, and yet many visitors do experience Dollywood as less contrived than other theme parks. On the surface such responses may appear odd, but, as this study shows, the park's success owes much to the development of the middle landscape over the course of the twentieth century. Moving from a pastoral landscape, the by-product of settlement, to the hyperreality of pastoral representations, the distinction between authenticity and inauthenticity seems to have become largely irrelevant.

Fig. 1: A replica of Dolly Parton's childhood home is displayed in the Dollywood theme park. Photo by author (2006).

In studying these developments I have chosen to focus on the Tennessee Valley. I have chosen this region in part because it was an area that I already knew from earlier research, but also because of the region's history. As a brief look at the settlement of the region will show, while the pastoral ideal was becoming increasingly popular in other parts of nineteenth century America, the rugged character of the landscape and its consequent isolation meant that the Tennessee Valley was only rarely included in that tradition.

Being geographically isolated, large areas in the Tennessee Valley remained economically, technologically, and culturally backwards compared to most of the nation (Schaffer, "Environment and TVA" 334). As British born geologist George William Featherstonhaugh noted in his journal when he travelled through the valley in 1830: "The people here, especially the women, were scarce above the level of savages, either in manner or appearance … they had neither butter, eggs, milk, nor vegetables of any kind. Salt pork, bad corn, bread mixed up with dirt, tobacco and whiskey, formed the whole list of their necessaries and luxuries" (Featherstonhaugh 173). Other travellers to the region also noted the delayed development. On his walk to the Gulf of Mexico, nature enthusiast John Muir wrote of eastern Tennessee and western North Carolina that it was "the most primitive country I have seen, primitive in everything. The remotest hidden parts of Wisconsin are far in advance of the mountain regions of Tennessee and North Carolina" (Muir 18). Little had changed thirty years later when the American forester Gifford Pinchot travelled the region in the 1890s. Pinchot's descriptions of mountain life also emphasized poverty, disease, and isolation, and even his description of small mountain farms is hardly the material of pastoral idyll: "[The Mountaineers] regarded this country as their country, their common. And that was not surprising, for they needed everything usable in it—pasture, fish, and game—to supplement the very meagre living they were able to scratch from the soil of their little clearings, which often were no clearings at all, but mere `deadenings,´ filled with the whitening skeletons of trees killed by girdling." Pinchot goes on to note the commonness of diseases such as hookworm, typhoid fever, and tuberculosis, and concludes that "the lives of these mountain people were literally as cribbed, cabined, and confined as the country in which they had their being was spacious, rich, and beautiful" (Pinchot, *Breaking New Ground* 61).

The poor living standards endured by many early Tennessee Valley settlers were to a large extent caused by the difficult and costly manner of transporting goods to and from the valley. To the north, east, and south the Appalachian and Cumberland mountain ranges complicated travels, and to the west, the wild and often unnavigable Tennessee River ensured that little trade was conducted via the waterways. Several attempts were initiated in order to domesticate the river for trade purposes, but all failed. Likewise, the construction of toll roads, such as the Saluda and Buncombe Turnpikes, constituted early attempts at creating passageways across the Southern Appalachians in order to break the isolation (Davis, *Where There Are Mountains* 127). But although such roads made crossing the mountains easier, they were often in poor condition and the cost of using them further raised prices on goods.

With few means of transportation, it was difficult for mountain farmers to get to market. As late as 1904 a Southern Appalachian mountaineer, explaining how the popularity of producing homemade whiskey was actually the result of isolation and poor roads, touched upon many of the realities of mountain life:

> [W]hen all's said and done, the main reason for this "moonshining," as you-uns calls it, is bad roads. … From hyar to the railroad is seventeen miles, with two mountains to cross; and you've seed that road! I recollect you-uns said every one o' them miles was a thousand rods long. Nobody's ever measured them, except by mountain man's foot-rule—big feet, and a long stride between 'em. Seven hundred pounds is all the load a good team can haul over that road, when the weather's good. Hit takes three days to make the round trip, less'n you break an axle, and then hit takes four. When you do git to the railroad, th'r ain't no town of a thousand people within fifty mile. Now us folks ain't even got wagons. Thar's only one serviceable wagon in this whole settlement, and you can't hire it without a team and driver, which is two dollars and a half a day. Whar one o' our leetle sleds can't go, we haffer pack on mule-back or tussle it on our own whethers. Look, then! The only far produce we-uns can sell is corn. You see for yourself that corn can't be shipped outen hyar. We can trade hit for store credit— that's all. Corn *juice* is about all we can tote arnd over the country and git cash money for. Why, man, that's the only way some folks has o' payin' their taxes! (Kephart 122-123).

Apart from the obstacles to trade, settlers in the Tennessee Valley also had to deal with a lack of available land. Toward the end of the eighteenth century the colonial government had given tracts of farmland to the men who had served in the French and Indian War, and by the early nineteenth century much of the remaining good farmland had been purchased by land speculators seeking to profit from the westward expansion. This left many independent farmers to settle on poorer lands in the mountains, and often in flood-prone areas.

When rivers and streams went over their banks, crops were destroyed and sometimes entire fields were washed away, forcing farmers to clear new land or move to other parts of the country.

Whereas parts of the Tennessee Valley were not always inviting to small farmers, the mountains, forests, and rivers did attract industry and often with severe environmental consequences. The very mountains that proved such a barrier to trade made possible a booming mining industry. The production of iron was among the earliest of such ventures that were established in the valley, and it left its marks on the landscape. In the words of historian Donald Edward Davis "by the third decade of the nineteenth century, the pounding of the great iron hammers could be heard echoing across the Southern Appalachians" (147). The charcoal needed for the iron production meant that large quantities of trees had to be cut down, a process that left barren and cut-over landscapes in its wake. Copper mining, which released a number of contaminants into the air that killed the surrounding vegetation, also required large quantities of wood, and gold mining often involved hosing away the hillsides with water cannons.

In addition to the different mining operations, the lumber trade played an important part in the region's economy. Early on small logging operations had targeted the highlands of eastern Tennessee, cutting down specific species such as oak and poplar. Because of the terrain it was almost impossible to establish permanent sawmills in these parts of the region, so teams of oxen were used to haul timber to the rivers where it was floated down to the lumberyards. By the late nineteenth century, improved roads, railroads, and the invention of the portable sawmill introduced widespread commercial logging to the area, but by then overcutting had already made its mark on the wooded hillsides.

As the history of mining and lumber industries in the region shows, the isolation of the Tennessee Valley did not keep inhabitants from participating in a thriving regional and national trade. In spite of the dangers of navigating the river, flatboats and keelboats carried crops and other produce to the markets in New Orleans; cattle and sheep roamed the forests and were often herded over the mountains to be sold at regional markets; and Tennessee was one of the largest producers of corn in the antebellum South, mostly due to the fertile bottomlands in the river valleys where the larger slave-driven plantations were located. Furthermore, farmers on small homesteads engaged in local trade that included crops such as nuts, fruits, berries, and honey.

If the isolation of the valley was not complete, its limited accessibility made the region less attractive to travellers. While tourists and artists praised and memorialized Niagara Falls, the Catskills, the Hudson River Valley, or the drama of the American West, fewer tourists went to the Great Smoky Mountains and no nationally acclaimed painters or writers portrayed the wonders of the Tennessee River. As Donald Davidson writes:

> One of the chief peculiarities of the old Tennessee is that, of all the great rivers east of the Mississippi, it has been the least friendly to civilization. Until the advent of the Tennessee Valley Authority it defied every human attempt at conquest. It could be used, but only at great hazard and on terms forbidding to commerce and industry. So it remained a wild river, cherishing its wildness while civilization rushed across it or away from it. It mocked the schemes of improvers. It wore out the patience of legislators. Tawny and unsubdued, an Indian among rivers, the old Tennessee threw back man's improvements in his face and went its own way, which was not the way of the white man. The white man therefore withheld his praise. So today there are no fine romantic songs about the Tennessee. Poets, novelists, makers of ballads and folk tales, all have passed it by. (*The Tennessee – The Old River* 6)

It was not until the 1880s, when Mary Noailles Murfree's stories popularized the Smoky Mountains, that people began to include the scenic wonders of that region in the romantic landscape tradition. By then, new and improved modes of transportation had also made it easier for visitors to get to the Southern Highlands.

By the late nineteenth and early twentieth century, modern versions of the pastoral began appearing in the design of public places in the Tennessee Valley. With the construction of the Biltmore mansion in Asheville, North Carolina, the idealized middle landscape was literally embodied in the structure of the estate. Likewise the preservation of the wilderness in the form of the Great Smoky Mountains National Park engaged a pastoral imagery. But nothing compared to the large-scale developments initiated by the Tennessee Valley Authority in the 1930s, when the valley was transformed in one of modern history's boldest and most ambitious demonstrations of public planning. However, the TVA's early pastoral vision would also signal an end to the era of major public works, and eventually the agency would play a central part in the move towards a series of increasingly decontextualized and empty pastoral representations.

This book traces and analyzes the development of the middle landscape from an unintended pastoral to a designed hyper-pastoral space. It is not meant to be an exhaustive catalog of the modern pastoral in the Tennessee Valley, but focuses on a limited number of places in order to posit a theoretical framework for further categorization, analysis, and discussion of the middle landscape's development after 1900.

The first chapter establishes the key elements of Leo Marx's middle landscape and ties the development of the pastoral to the simulacrum theory of the French postmodernist Jean Baudrillard. His analysis of how reality is simulated to the point where the copy supersedes the original is relevant to the development of the pastoral which enters first a phase of pastoralization, then a phase of mass-production during which it is freed from its original context, and thirdly a state of hyperreality where the pastoral no longer has any referentiality. The last part of the chapter analyzes the Museum of Appalachia as an example of the pastoralized space and the politics of historical representation.

Chapter two looks at three early copies of the pastoralized space: George Washington Vanderbilt's impressive Biltmore Estate, the Appalachian Trail, and the Great Smoky Mountains National Park. These are among the first examples of how the pastoral was transplanted into a contemporary context. Vanderbilt's castle, for example, was modelled after several existing French estates and included landscaped gardens, working farms, and America's first scientifically run forest. Likewise, the pastoral is central to the plans for an Appalachian Trail and the creation of the Great Smoky Mountains National Park.

The third chapter is a study of the Tennessee Valley Authority's reliance on pastoral ideology as it transformed the region. Created primarily to raise the standard of living and to break the economic cycle of the Depression, between 1933 and 1942 the TVA initiated a number of projects that were both implicitly and explicitly pastoral. Looking at some of the initiatives in the Norris Dam area—reforestation, the transformation of farmland into pastures and orchards, the development of recreational resources, and the construction of the experimental town of Norris—the chapter details the transformation of the valley into a designed middle landscape. The last part of the chapter looks at how the TVA's middle landscape, in spite of its socially egalitarian appearance, marginalized and excluded black Americans as well as the large houseboating communities who called the river their home.

Chapter four details the breakdown of the pastoral ideologies that had guided many of the earlier projects. In constructing the Tellico Dam and the Tennessee-Tombigbee Waterway the pastoral was freed from its context and meaning. Both Tellico and the Tenn-Tom represented large-scale post-war public works projects undertaken primarily for economic reasons. Yet, as public opinion turned against the projects, threatening their continued funding, recreational and environmental issues were used in order to justify them.

Chapter five focuses on a group of modern tourist attractions that use the pastoral as their central narrative metaphor, but which do so in a nostalgic, fragmented, and decontextualized form. At the Chattanooga Choo Choo, an old train station restored and transformed into a hotel and restaurant complex, an antique train is displayed amidst a blooming gardenscape, allowing visitors not only a unique glance into a highly stylized past, but also an opportunity to spend the night in one of the refashioned antique train cars. At Dollywood the story of singer Dolly Parton's romanticized Smoky Mountain childhood is used in order to infuse the popular attraction with a sense of history and authenticity. Displaying traditional arts and crafts, along with antique technologies and a replica of the log cabin Parton lived in as a child, the theme park is selling an idealized version of the past. And in Lynchburg, Tennessee, the Jack Daniel's distillery is simultaneously the town's largest employer and tourist attraction. Since the 1950s the brand has relied on a PR campaign that highlights local residents and promotes idealized small-town values. The resulting blend of marketing strategy and communal identity makes it impossible to identify the line between the authentic Lynchburg and what might best be described as "the Jack Daniel's experience".

Looking at the history and narrative structure of a number of different places in the Tennessee Valley, this study traces the development of the modern middle landscape as it moves through a process of simulation. In that process the structure of such places reveals not only an increasing turn towards nostalgic representations designed to meet public demands, but also how expectations of nature have changed over time as notions of *the real* are replaced by a staged authenticity. This book looks at the development of the pastoral and how it has been adapted to fit the modern world, simultaneously reflecting the surrounding society and offering alternatives to it. To this day, the pastoral remains central to the American experience.

One

The Museum of Appalachia and the Pastoral Simulacrum

I

When Leo Marx coined the term "middle landscape" in 1964 he sought to describe a prominent trend in nineteenth century American literature; a clash between the idea of the pristine landscape as a sacred national heritage and the inherent potential for progress that was associated with the vast and, supposedly, empty continent. As both a real and an artistic space where the machine and the garden met, the middle landscape became the setting where Americans sought to reconcile those two ideals.

In many ways the middle landscape was a peculiarly American version of the pastoral, a literary tradition that was first popularized by the Roman writer Virgil in the first century B.C. Virgil's fictional Arcadia served as a refuge from the political world of Rome. In the open and green countryside that would become the archetypal pastoral setting, man found an escape from the corruptions of the city without having to face the dangers of the wilderness. In that sense, the pastoral landscape essentially coupled the primitivism of the natural world with civilization's desire to control. Thus, even as early as Virgil's work the pastoral was a moral space that seemed to offer a social alternative. The Virgilian countryside was a landscape infused with a sense of cultural primitivism, and the people living there, the pastoral seemed to suggest, were endowed with a distinct sense of nobility and piety. The pastoral hero, often a shepherd, and the protagonist of the related agrarian *Georgics*, having retreated into the countryside, were no less civilized than their urban neighbors but were "in tune with nature to the extent of acquiring superior wisdom and insight" (Rowe 219).

The pastoral remained among the most prominent nature ideals in the Western world, and held such sway over the European imagination during the age of discovery that, according to Marx, it came to embody America itself: "The ruling motive of the good shepherd, leading figure of the classic, Virgilian mode, was to withdraw from the great world and begin a new life in a fresh, green landscape. And now here was a virgin continent! Inevitably the European mind was dazzled by the prospect. With an unspoiled hemisphere in view it seemed that mankind actually might realize what had been thought a poetic fantasy" (3). It is debatable whether the American fascination with the pastoral should be attributed to a neo-classical focus on recreating Arcadia, as Marx argues, or if, as Charles Sanford has noted in his review of *The Machine in the Garden*, the model for American pastoralism was in fact "not Arcadia, but the Garden of Eden." Although the Classical Revival was of great importance to the formation of the American character, in the end, Sanford writes, the influence of the Graeco-Roman heritage was "less important and less enduring ... than that of the Protestant Reformation" (Sanford 275). What both Marx and Sanford agree on, however, is that the pastoral vision was so integral to the American experience that it became central to the formulation of a political and cultural framework for the young nation.

But in spite of Thomas Jefferson's famous declaration that America should remain a democracy of yeoman farmers and that all factories—along with their corrupting influences on the soul of man—should stay in Europe, by the nineteenth century the industrial revolution was rapidly transforming the American landscape. Adapted to fit this new experience the American pastoral gave way to the middle landscape. In many respects the middle landscape was derived from and similar to the pastoral in that it sought "a resolution of the opposed worlds of nature and art" (Marx 22). But unlike the simple structure of the pastoral that unproblematically offered an escape from modernity, the middle landscape harbored a sense of conflict between the longing for progress and the admiration for a landscape that many saw as a divine creation that granted America its claim to national exceptionalism. The middle landscape was, Marx suggested, "a new, distinctively American, post-romantic, industrial version of the pastoral design" (32).

The middle landscape soon became a popular narrative because it reconciled man and nature in a way that seemed to suggest that neither was complete in itself. In the words of Patricia Greiner, the middle landscape served as a "vision of moderation … [a] midway between untrammelled wilderness and artificial urbanity … characterized by peacefulness and seclusion in comparison to the city, by order and certainty in contrast to the primitive wilderness, and above all by usefulness" (8). The middle landscape afforded Americans a way of maintaining the pastoral vision of both social and environmental control in an industrialized world.

The history of the scenic Hudson River Valley is a good example of how the middle landscape included the theme of industrialization but narrated its way out of the social and environmental implications. Famous for inspiring a school of landscape painters, by 1850 the Hudson Valley had become a largely engineered environment. Early in the nineteenth century the Hudson River was already lined with industries, and steamboats regularly plied its waters with tourists, settlers going west, and products being brought to and from markets. Before 1830 the region had seen the completion of two major man-made waterways: the Erie Canal in 1825 and the Delaware-Hudson Canal in 1828. By the middle of the century the natural bends of the river were being straightened and adjoining wet lands filled in to accommodate the railroads. In only fifty years, the formerly wild and beautiful Hudson Valley had become industrialized to the extent that it was damaging the area's scenic appeal.

However, none of these developments deterred painters such as Thomas Doughty, William Barlett, and Thomas Cole from painting scenic Hudson Valley landscapes. Barlett's paintings in particular, focusing on boats and buildings, show the river as part of an idyllic landscape that was dominated, but unthreatened, by civilization. Doughty and Cole also painted landscapes that were often mediated by a solitary man fishing or wandering in the landscape, or a steam locomotive disappearing into the lush countryside. While thus acknowledging the machine's intrusion into the "garden," the middle landscape narrative allowed Americans to believe in a peaceful merger between nature and culture by ignoring the more disturbing accounts of industrial exploitation and environmental conquest. As the painter Thomas Moran once told a critic who questioned the accuracy of his paintings: "I place no value upon literal transcripts from Nature. My general scope is not realistic, all my tendencies are toward idealization …. Topography in art is valueless … and, while I desired to tell truly of Nature, I did not wish to realize the scene literally, but to preserve and to convey its true impression" (Anderson, "Curious Historical" 16).

The middle landscape narrative allowed the pastoral ideal to survive the industrial revolution, albeit in a modern version, and by the twentieth century the pastoral ideal still permeated American society. At this point, it might be relevant to clarify what I mean by the term *pastoral* since it is a rather central term for this study. Terry Gifford has defined three versions of the pastoral: as a historical and literary form rooted in centuries of poetic use, a broader use that describes a space "with an implicit or explicit contrast to the urban," and finally a more critical use of the term as too idealized to be useful in a modern environmental context (1-2). The use of the term in this study adheres to the second version, and more broadly includes landscapes that tend to be positioned in opposition to the strictly urban. In that sense, the modern pastoral no longer necessarily revolves around a shepherd, or a small cabin with a patch of land cleared for farming, but rather comprises the more general idea of retreating into the countryside, just as it remains associated with a sense of simple nobility, wisdom, and even authenticity. The modern pastoral may include recreational areas, modern suburbs, and the design of the American college campus. It might be a relevant term in discussing the architectural style of a Frank Lloyd Wright, who revolutionized suburbia by creating a new organic home that preached a high level of correspondence between the house and its natural surroundings. It might also prove useful in a discussion of the Levittown complex in Pennsylvania, where in 1950 Levitt and Sons, Inc. created a suburban community that would eventually house over 60,000 residents, and as part of that project set aside a 250-acre forest reserve for recreational purposes, and planted approximately 48,000 blooming fruit trees within the suburb itself (Rowe 50).

Tracking the development of the pastoral from its simple to its complex form, Marx offered a valuable insight into the metaphysical make-up of the American nation, but he never further developed his theory by bringing the middle landscape into the twentieth century or by nuancing it. Howard Segal has usefully divided the middle landscape into three sub-categories: the urban middle landscape which seeks to balance the city against the surrounding countryside, the suburban which, at least to some extent, denotes the desire to escape the city, and the regional which often revolves around efforts to conserve natural resources. This study is meant as an attempt to trace and describe the development of the middle landscape in the twentieth century, by looking more exclusively at the Tennessee Valley.

II

In 1981 the French postmodernist and cultural theorist Jean Baudrillard formulated his theory of the simulacrum, or the copy without an original. According to Baudrillard the crisis of the modern is caused by the substitution of empty signifiers for reality. Hiding the lack of reality behind a simulation of the real, the traditional frame of referentiality has been broken, creating instead a hyperrealist mode in which the two poles of signifier and signified implode and become one. As a consequence, the only reality to which the copy eventually refers is its own.

In his study Baudrillard distinguishes between three distinct phases of the simulation process. The first phase concerns the counterfeit, which essentially reproduces the original in a form that is still identifiable as a copy. The second order of simulacra describes the repetitive and systematic mass production of signs which works to free the sign from its specific context; and in the third order of the simulacrum reality is reduced to pure simulation.

Although I do not mean to suggest that the development of the middle landscape complies with Baudrillard's theory of the simulacrum in detail, the movement towards a copy without an original is in many ways indicative of the modern pastoral's development. The process of simulation presupposes that any reproduction happens according to a design. In this case then, the simulation of the pastoral necessitates a move from an *unintended* to a specifically *designed* type of space. The unintended middle landscapes, such as eighteenth century New England farms, constitute the original, to the extent that it is possible to talk about an *original* pastoral space, whereas the designed spaces—the objects of this study—form the procession of copies leading to the eventual disappearance of the original.

Marx's complex pastoral might best be described as an unintended middle landscape as embodied in texts. It was a by-product of the nation's struggle to come to terms with the transition from an agrarian to an industrial society, just as the eighteenth century

New England countryside, for example, developed as the result of individual farmers trying to carve a living out of the forest. More recent examples of an unintended landscape include, as David Nye shows in *American Technological Sublime*, the famed New York skyline which is the result of individual buildings, each built without regard for the general pattern, forming an aesthetic whole with the blue of water and sky (90). As this example also suggests, even though there is a general chronological development from the unintended to the designed, the unintended middle landscape is still produced on a regular basis, often coexisting with the designed space.

As the twentieth century progressed, the middle landscape increasingly became a designed space, directed at creating a particular kind of experience. Urban parks and open-air museums are examples of how places were specifically designed to generate a certain response, but even at the shopping mall, nature—in the form of fountains or potted trees and plants—is often used to create a relaxed and comfortable space to which customers will want to return (Farrel 33-37).

But merely dividing the middle landscape into an unintended and a designed space does not do justice to the historic complexity of its development. It is possible to identify four subordinate categories of the middle landscape. The unintended middle landscape can be divided into two phases: the *pastoral* and the *mechanized pastoral*. While the pastoral denotes the original bucolic experience—a small farm with fields, a barn, an orchard, and some animals—the mechanized pastoral describes the industrialized version of that space. Over the course of the nineteenth century, new factories often were located in scenic surroundings in order to benefit from the natural resources. Rivers afforded industries both cheap and plentiful power and a means of waste disposal. In fact, so popular was the Niagara Falls as an industrial site that by the 1860s industrial use diminished the flow of water over the falls to an extent where the natural wonder no longer lived up to tourists' expectations (Spirn 95).

Likewise, the designed space can be divided into *pastoralized* and *hyper-pastoral* phases, a development which adheres to the Baudrillardian process of simulation, moving from counterfeit to hyperreality. In the early part of the twentieth century, the post-frontier crisis in American society combined with reform-minded progressivism to create an imagined pastoral ideal in new and often utopian contexts: the pastoralized space. By copying the earlier unintended pastoral into a designed contemporary context, projects such as the Biltmore Estate, the Appalachian Trail, and many of the TVA's early programs exemplified this new kind of place that essentially counterfeited the pastoral. Based in part on actual pastoral experience, and in part on the imagined values that were associated with it as an artistic ideal, the pastoralized place often served as a moral landscape, supposedly suggestive of a Jeffersonian, pre-industrial and non-exploitative social alternative.

The ideology connected to the earlier pastoralized space broke down as the pastoral design was mass-produced. With the creation of large-scale public works such as the Tellico Dam and the Tennessee-Tombigbee Waterway, the middle landscape became a means to justify projects that were originally economic in nature, but which did not seem acceptable to the public. Freeing the pastoral from its context, the breakdown of ideology signaled a move towards the hyper-reality of places such as Dollywood and the Jack Daniel's distillery, where the copy seems to have superseded "*the real*" to the extent that it is no longer possible to draw a line between the two.

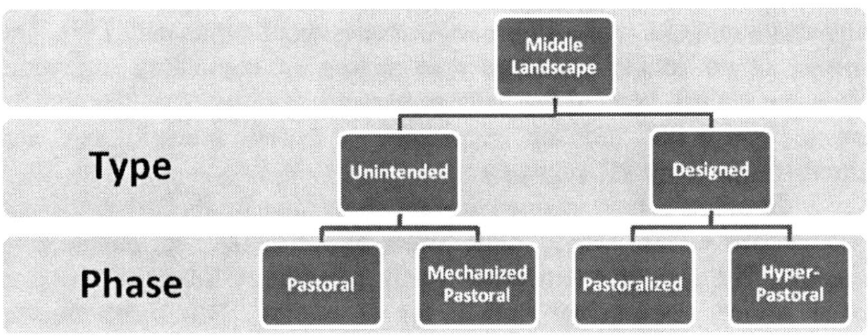

The experience of authenticity is central to the appeal of the pastoral, but as Baudrillard's theory suggests, in a world dominated by simulation it has become increasingly difficult to experience anything truly genuine. Part of American culture, as Umberto Eco has argued, is obsessed with "reality," a fascination which ironically has filled the nation with realistic copies or markers of past events (4). But if demand for the real thing necessitates the creation of "the absolute fake," as Eco writes, the boundaries between reality and unreality become blurred (8). A case in point is media mogul William Randolph Hearst's Hearst Castle where, among other curiosities, historic buildings, brought in and shipped from Europe, were reassembled among complete fabrications in the creation of a historicized, and ultimately hyperreal, space (21-22). But the example of Hearst Castle also forces one to wonder, what happens when that which is reproduced never existed? Does the representation lose its sense of authenticity and thereby its appeal, or does it even matter?

The question about the extent to which it is even possible to experience something authentic in a consumer-driven culture where every site and experience is capitalized upon, has been debated since 1961 when Daniel Boorstin published his groundbreaking book *The Image*. Boorstin argues that the increasing focus on advertising along with the media's growing tendency to create news rather than to merely report them has caused expectations of life to become so overpowering and unrealistic that people increasingly gravitate towards events that are planned and orchestrated in order to fulfil their exaggerated demands. Tourists increasingly expect "that we can make the exotic an everyday experience (without its ceasing to be exotic); and can somehow make commonplaceness itself disappear" (77). The tourist is no longer necessarily in search of something authentic, Boorstin claims. Instead, he isolates himself from the societies that he visits through the uniform experience of hotels, guided tours, and ritualistic patterns of behaviour.

Boorstin's book gained instant fame due to its insightful and provocative ideas, which anticipated hyperrealist and postmodern thought. His pseudo-event certainly fit Baudrillard's later description of a society increasingly made up of copies. The focus on the plasticity of modern society spread into other academic disciplines as well. In 1976 geographer Edward Relph used the notions of authenticity and inauthenticity to study place in modernity. "Authenticity," Relph writes, "connotes that which is genuine, unadulterated, without hypocrisy, and honest to itself not just in terms of superficial characteristics, but at depth" (64). He goes on to show that inauthenticity in fact defines a majority of modern places, giving examples such as suburban landscapes, highways, tourist attractions such as theme parks, and recreated or restored pioneer villages which engage in a process of museumization by displaying the past in a temporal, social, and political vacuum (101-103).

In his anthropological study *The Tourist*, published the same year as Relph's book, Dean MacCannell answered Boorstin's critique by arguing that the very idea of a pseudo-event/reality dichotomy suggests the existence of something real and at the same time reduces the inauthentic to something transitory. MacCannell sees the in-authentic as far more permanent and argues that the tourist is in fact a representative modern man in search of something authentic in a world void of meaning but full of surfaces and empty signifiers. If the authentic is non-existent, or at least rare, modern society no longer has any use for such distinctions as authenticity and inauthenticity. MacCannell sees the tourist in opposition to the postmodernist who embraces the "historyless void" of endless representation (xii). The tourist is still searching for meaning in a historical context, even if that context is reconstructed. But if the boundaries between the authentic and the inauthentic have broken down, how can one make that distinction? To account for this dilemma, MacCannell introduces the notion of staged authenticity.

According to MacCannell, the sense of authenticity is often related to exotic cultures, nature, and representations of history. But a sense of the authentic can also be found merely by watching the "real" life of others. Modern man "is losing his attachments to the work bench, the neighbourhood, the town, the family...but, at the same time, he is developing an interest in the 'real life' of others" (91). For those reasons, modern tourist attractions have increasingly staged such displays in order to sell a sense of authenticity to the tourist. As later chapters will show, handicraft displays, for example, have remained popular for those very reasons: the exhibitions of craftsmanship function both as signifiers of a historically loaded and exotic anti-modernity, while simultaneously allowing spectators to view an imagined everyday experience. Likewise, allowing visitors to watch food being prepared in a restaurant or to take a behind-the-scenes factory tour may serve to heighten the visitor's feeling that he is witnessing something real.

The authenticity debate is relevant when looking at the development of the middle landscape, since the pastoral has always appealed to people as an anti-modern response to progress. Yet the process of simulation has increasingly made the pastoral a part of modernity. In this sense, the hyper-pastoral sells an experience of anti-modernity without actually affording the visitor any escape from inauthenticity; and as the popularity of places like Dollywood shows, visitors readily accept the new escapism of the hyper-pastoral.

In his book *Making a Middle Landscape* Peter G. Rowe argues that even as part of modernity the middle landscape is a potential answer to the problem of placelessness. Rowe's study of the pastoralism of suburban and ex-urban developments in modern America looks closely at four different middle landscapes: single family homes, retail facilities, office parks, and highways. As a synthesis between pastoral myths and modern technology, Rowe's "modern pastoral" harbors the same anxieties as Marx's middle landscape, and fails to bring about the sense of order and structure that is inherent in the pastoral. According to Rowe such problems mainly stem from the fact that the merger of the pastoral's morality with the suggestion of mass-production and efficiency has remained asymmetric and oppositional (232-234).

The modern middle landscape may be an oppositional structure at heart, but through the right kind of poetics it can come to function as an antidote to the placelessness of modernity. Modern places, Rowe argues, express a "sameness, predictability, and blandness [that] either assaults through blaring signage and light or stealthily confronts through anonymity and blankness. Either way—roar or silence—the landscape appears to be undifferentiated and lacking in any real meaning, let alone profundity" (Rowe 56). In opposition to French anthropologist Marc Augé's idea that the age calls for an anthropology of non-places—since we spend so much of our lives on freeways, in airports, and in front of ATM machines—Rowe somewhat optimistically argues that through the pastoral design we may eventually correct the feeling of not belonging that is such an integral part of modernity. Whether Rowe's optimism is sound only time can tell, but if MacCannell is right in his claim that there is little or nothing that distinguishes the authentic from the inauthentic in the modern world, then the designed and mass-produced pastoral simulation may be worth more than its mere commercial appeal suggests.

III

John Rice Irwin's Museum of Appalachia is located on the outskirts of the small town of Norris, Tennessee, a mere half hour drive from the hustle and bustle of Knoxville. The museum, which was turned over to a non-profit organization in 2003, began as a small private collection of Americana with the purpose of preserving regional heritage and culture. At present it has become a major local tourist attraction, drawing thousands of visitors every year. In 2006 it was awarded a prize for historic preservation by the Tennessee Society of the Daughters of the American Revolution, and the following year it was made an official affiliate of the Smithsonian ("The Museum of Appalachia named"). Over the years the museum has also been portrayed in as diverse media as the *Reader's Digest*, a centerfold article in *Parade* magazine, and before a national audience on NBC's "Today Show" (Irwin, *The Museum of Appalachia Story* 5).

The Museum of Appalachia is an interesting example of a first order pastoral simulation. Its success is, at least partly, due to its narrative structure. Portraying the regional past as a simpler and more harmonious time, it appeals to modern cultural nostalgia by offering visitors a chance to escape, albeit briefly, from everyday life. Already upon reading the museum's brochures, or visiting its website, one suspects a sentimentalist approach to history, and walking about the museum grounds among the impressive collection of log cabins that suspicion is soon confirmed. Irwin's museum fits in well with the larger American discourse of hardy pioneers creating a garden landscape out of the wilderness. The tourist is surrounded by pastures and split rail fences, wandering peacocks, grazing sheep, and live country music.

Fig. 2: The Museum of Appalachia in Norris, Tennessee. Photo by author (2007).

Irwin's museum is more than mere nostalgia; it is a moral landscape that laments the disappearance of a culture in the face of modernity. As Irwin describes it in one of the museum brochures: "I hold the strong conviction that the true breed of diminishing mountain folk of Southern Appalachia are among the most admirable people in the world. They are kind, gentle and compassionate, yet tough and resourceful" (Irwin, "Museum of Appalachia: A living Mountain Village"). The brochure conjures up an image of the pioneer farmer as a rapidly disappearing American archetype. Just as the open green valley of the museum tells a tale of domesticating the wilderness, so a handwritten sign next to the display of the American Axe reads: "Our most important tool, without which the howling wilderness would have so remained." In his book *The Museum of Appalachia Story* Irwin further explains that the axe was the tool "responsible for converting the wilderness into the tranquil, domestic and agricultural countryside capable of feeding and clothing a growing and vibrant nation" (30). Using the Puritan rhetoric of William Bradford's "howling wilderness," Irwin's axe exhibit evokes an epic story of nation building, and further suggests the age-old narrative of conquering the wilderness to create a divine civilization, of creating a moral landscape from an immoral one. It is a version of American history that is highly relevant to understanding the museum itself.

The anti-modernity theme found at the Museum of Appalachia is common to the open-air museum. Skansen—a replica of a nineteenth century Swedish town traditionally hailed as the first open-air museum—was created by Swedish scholar Artur Hazelius in Stockholm in 1891 as a response to the rapid industrialization of the times. Hazelius, who was interested in peasant culture, collected well over a hundred historic buildings and placed them in a bucolic setting as a tribute to the country's rural past. The museum eventually proved a major source of inspiration, and in the following decades, open-air museums began appearing all over Scandinavia, the rest of Europe, and America.

This inherent critique of industrial society carried over into early twentieth century American heritage centers such as Henry Ford's Greenfield Village and John D. Rockefeller Jr.'s Colonial Williamsburg. Somewhat ironically, considering how Ford and Rockefeller had made their fortunes, the two museums both sought to recreate a pre-industrial experience, imitating the pastoral by establishing a retreat from the industrialized and morally ambiguous modernity; both Greenfield Village and Colonial Williamsburg are idealized versions of a fresh start in a landscape that retains its original character while under human control.

Ford's idea to create a museum originally came about when his childhood home in Dearborn, Michigan, was threatened by road construction. To save the building, Ford bought it, moved it, and undertook to renovate it to match his childhood memories. In the following years he collected other old buildings, either because they matched his memories of the town in which he grew up, or because they had belonged to historic Americans that he admired. It was clear from the start that Ford's vision was a highly personal and idealized version of his hometown with added features relating to a more general American history. When completed, Greenfield Village had become an idealized small town complete with a white-spired church, a courthouse, a school, and an inn all set about a New England style village green. Ford also revived folk dances and traditional dresses.

Visitors could see history played out before their eyes in the form of old buildings, demonstrations of crafts, and collections of Americana, but at the same time they could see a stylized version of a golden-age America, simultaneously a symbol of what had been lost and what the nation might become again.

From a historian's point of view the pastoral structure is far from unproblematic. Escapist at heart, places such as Greenfield Village, Colonial Williamsburg, and the Museum of Appalachia essentially offer visitors a chance to retreat from modernity into a simplified world that embodies everything that modernity is not. Built around an idealized vision of a world devoid of social and political problems, visitors who enter that spatial and temporal vacuum, escape not only the modern but history itself.

Just as in Greenfield Village or Colonial Williamsburg, visitors to the Museum of Appalachia encounter a nostalgically driven narrative of Edenic simplicity, harmony, and bliss that emphasizes the life and values of the white, independent farmer of implicitly English or Scots-Irish background. Irwin cultivates what amounts to an idealized pre-capitalist, pre-industrial, and essentially pre-lapsarian America frozen in time. The cabins, churches, and tools all tell stories of good-natured and self-reliant mountain folk without important ties to a regional or national market. In some respects, the story that Irwin tells evokes a Jeffersonian vision of a non-industrial nation based on the virtues of the hardworking, democratic, and honest yeoman farmer whose knowledge of, and connection to, nature is at the very heart of the American character.

> [My grandfather] had known many of the old pioneers and he had learned from them the secret of the forests, the streams, and the countryside. He taught us the names of every plant, every vine, and every tree. He pointed out that the ash split easily and made good stovewood, that the hickory made good handles for axes and hammers, that the cedar and locust made good fence posts, and that the chestnut made good rails. He showed us, deep in the woods, where the muscadines grew, and where the huckleberries, the `possum grapes, and fox grapes could be found. He took us to the top of Pine Mountain one

> time and showed us where the summer grapes grew. (Irwin, *The Museum of Appalachia Story* 14)

Irwin's childhood memories, combined with a larger American discourse of trail blazers and pioneer farmers who tamed the wilderness and paved the way for civilization, form the museum's narrative structure.

However, just as Jefferson's vision was not viable in early post-revolutionary America, so the pervasive image of the lush, green museum grounds hardly does justice to an Appalachian past that was as hard, brutal, and lonely as it was complex. And it is an image of idyll which visitors take from the museum, even if Irwin has not attempted to hide the harsh realities of mountain living. As one visitor noted, she could easily imagine an entire family living in the cramped space of the small cabins, the "long days, bending over stone fireplaces for hours to cook the meals," and the uncomfortable ruggedness of the small mountain church's seating arrangements. Yet in spite of the imagined hardships, the experience brought her "comfort and calm, and renewed energy" as she sat on the porch of one cabin gazing over "the cow pasture outlined with the split-rail fence; the garden where the corn was turning yellow and the peppers red in anticipation of the harvest … the neat patches of tobacco plants amid the green lawn; and the other cabins, mill, blacksmith shop and cantilevered barn I could see off in the distance." Vowing to be back to visit the museum again, it would still be in that "wide green valley" where most of her time would be spent (Reuter).

Visiting the Museum of Appalachia is indeed a pleasant and informative experience, and the work which Irwin has put into recreating the past as he remembers and has experienced it is both commendable and honest; certainly there is no guarantee that a trained historian would have structured a museum of the Appalachian past any differently. And the pastoral narrative was a central part of life in the mountains. Large tracts of mountain land were cleared for farming, and social events such as cornshuckings meant that entire communities gathered in order to celebrate the harvest and the passage of life (Davis, *Where There Are Mountains* 137-138). Horticultural practices, including the harvesting of nuts, fruits, and berries, were also an important part of the subsistence culture of the region, as was the production of honey. But as with any exhibition, the Museum of

Appalachia narrative inadvertently transforms history by what it includes and leaves out, by how the exhibits are spatially and visually composed, and by the kinds of information that the visitor is given.

Fig. 3: The Mark Twain family cabin displayed at the Museum of Appalachia. Photo by author (2006).

One example of the museum's narrative content is the absence of the TVA from the exhibits. Although the agency was instrumental in creating much of the pastoral framework that is so central to Irwin's museum, and although the construction of Norris Dam flooded his family's ancestral lands forcing them to sell their farm to the government when Irwin was only four years old, the TVA is not a part of the history presented at the Museum of Appalachia. Large federal subsidies and government intervention, it seems, hardly fit a tale of independent mountaineers. If some issues are completely left out of the narrative, others are merely given a positive spin. Poverty, for example, is mentioned only to accentuate the generosity of the mountain folk who willingly shared what little they had, and there are no displays pointing to the many diseases that were common to the area, such as pellagra and hookworm, which were caused by an inadequate diet, dirt floors, and going without shoes. Malaria also posed a serious threat to people at the time, affecting, according to Centers for Disease Control estimates, close to one third of the region's population prior to the arrival of the TVA.

Likewise, the displays that point to the industrial interests of lumber and mining companies in the region are never truly associated with issues of politics, race, or class, and the stories of large-scale environmental destruction caused by various industrial projects that left in their wake a landscape of clear-cut hillsides, low soil fertility, and desert-type environments caused by the release of airborne contaminants, are also absent from mining exhibits. Indeed, there is a major discrepancy between the story told by the lush and open spaces of the Museum of Appalachia experience and Thomas Wolfe's harrowing descriptions of a Smoky Mountains landscape at the turn of the century.

> The great mountain slopes and forests of the section had been ruinously detimbered; the farm-soil on hill sides had eroded and washed down; high up, upon the hills, one saw the raw scars of old mica pits, the dump heaps of deserted mines. … It was evident that a huge compulsive greed had been at work: the whole region had been sucked and gutted, milked dry, denuded of its rich primeval treasures: something blind and ruthless had been here, grasped and gone. (Wolfe 236-237)

In many places the Smokies were no longer the landscape of legend and myth, and although the past still existed in the exaggerated accounts of yesterday and in the form of poor mountain farmers in small shacks hanging on to memories of place and family, "their inheritance was bare. Something had come into the wilderness, and had left the barren land" (Wolfe 237).

Irwin's museum does include exhibits that potentially could counter the dominant narrative, but these stories are never allowed to question the validity of the pastoral structure. The displays of Native American history and culture, for example, focus on portraying the tribes of the area as parents and farmers, showcasing everyday artefacts such as clothes, woven baskets, and tools. But although Irwin's attempt to humanize the tribes of the region is admirable and historically correct, there is nothing pointing to the horrors of forced relocation necessitated by the process of creating a garden in the wilderness which the axe exhibit celebrates. Likewise, there are no displays pointing to the significant presence of African American workers in the iron industry of the East Tennessee Mountains. Indeed, the museum only recently added a slave cabin to its collection.

In spite of the obvious problems of achieving historical accuracy within a narrative framework, claims of authenticity are nevertheless central to the appeal of places such as the Museum of Appalachia, as Irwin is well aware. In a museum brochure, he explains that he has "aimed for the `lived-in´ look, striving for, above all else, authenticity [and] to make the Bunch House, the Arnwine Cabin, and all the other dwellings appear as though the family had just strolled down to the spring to fetch the daily supply of water" (Irwin, "Museum of Appalachia: A living Mountain Village"). All exhibits are presented with a detailed description placing them in a very specific context, explaining what the item was originally used for, as well as who owned or made it, a practice born out of Irwin's belief that if "an item is separated from its history, then its meaning and significance is greatly diminished, or lost altogether" (Irwin, *The Museum of Appalachia Story* 23).

The Museum of Appalachia is a wonderful example of the pastoralized landscape. Much like Greenfield Village and Colonial Williamsburg, Irwin's museum is a counterfeit, a recognizable copy that refers back to the immediate reality of the people whose lives Irwin is trying to memorialize, and to the experiences of his own childhood in the mountains of East Tennessee. Celebrating the past by selling visitors an anti-modern experience and a sense of authenticity, Irwin's museum represents a first step in the development toward the disappearance of the distinction between the real and the staged. In fact, although Irwin and his partners have put much effort into the preservation of the past, one wonders if, for many twenty-first century visitors, it makes any difference that the museum is not full of replicas.

Two

Estates, Trails, and National Parks: Early Examples of the Pastoralized Space

I

The geographic isolation that proved such a determining force in the settlement of the Tennessee Valley meant that for much of the nineteenth century the region received little attention from tourists and artists. In other parts of the country the Transcendentalists were drawing attention to the wonders of the New England landscapes, and the painters of the Hudson River School had already established the Hudson Valley, the Catskills, and Niagara Falls as areas of particular scenic interest. By mid-century, as the expansionist gaze focused on the western territories, places like the Sierra Nevada and the Yellowstone region took hold of the national imagination. Yet only few Americans looked to the Southern Appalachians for recreation or artistic inspiration. No famous painters depicted the natural wonders of the Smoky Mountains and there was no Mark Twain to celebrate the charms of the Tennessee River and its scenic tributaries.

That is not to say that there was no tourism in the region before the Civil War. In the early nineteenth century western North Carolina was a preferred summer home for many of South Carolina's well-to-do planter families in their efforts to avoid the deadly summer fevers that thrived in the mosquito filled humidity of the lower lying regions. Only the wealthiest part of the population could afford to keep such a summer home, but for those who could afford it, the cooler and drier mountain climate became a refuge (Starnes 13). Throughout much of the century various hot springs all over the South had also drawn crowds from all over the eastern seaboard, until eventually entrepreneurs began marketing the clean mountain air and the mineral springs as having a healing effect on a variety of illnesses. Before long hotels and shelters began appearing in the foothills of Southern

Appalachia, providing accommodation for visitors looking to be cured from respiratory ailments like tuberculosis (Brown, *The Wild East* 80).

In spite of such early tourist trends, it was not until the 1870s and 1880s when local writer Mary Noailles Murfree began publishing her stories about life in the Smokies that the Tennessee Valley was included in the romantic tradition that had drawn people to so many other scenic parts of the country. Born on a farm in Murfreesboro, Tennessee, in 1850, Murfree was partially paralyzed by an illness early in life. Spending her summers in the mountains, it was her love of the area and her admiration for contemporary local color writers like Bret Harte and Mark Twain that gave her the idea to write of the land that she knew so well. After the Civil War she made arrangements to live with a family in Cades Cove, and started travelling the mountains on horseback looking for scenery and inspiration for her stories.

Murfree's stories were first published in 1874 in *Appleton's Journal* under the pseudonym Charles Egbert Craddock, and by 1878 she was getting published in the *Atlantic Monthly* as well. Her stories combined drama, tragedy, and romance, helping the reader to envision Southern Appalachia as a romanticized alterity and the people living there as exotic Others. Her picturesque landscape descriptions were so vivid that the contemporary reader soon became ensnared:

> The wild grapes were blooming. Their fragrance, so delicate yet so pervasive, suggested some exquisite unseen presence—the dryads were surely abroad! The beech-trees stretched down their silver branches and green shadows. Through rifts in the foliage shimmered glimpses of a vast array of sunny parallel mountains, converging and converging, till they seemed to meet far away in one long level line, so ideally blue that it looked less like earth than heaven. ("Drifting" 6-7)

Such descriptions inspired an urban population used to long hours of hard toil in factories and offices, and Murfree's novels became instrumental in drawing people to the Smoky Mountains. As Allen W. Batteau notes, it was with the local color movement and "primarily through the writings of Mary N. Murfree, that Appalachia became a national reality" (39).

Many of Murfree's protagonists were outsiders who discovered, and were influenced by, the wonders of the Appalachian landscapes. In the short story "The Romance of Sunrise Rock," for example, it is the intrinsic freedom of the mountains that convinces the two central characters, Trelawney and Cleaver, to leave the city behind. Having otherwise failed in their ambitions the two young doctors embark on the self-imposed exile in the Smokies, but while Trelawney immediately surrenders to the new way of life, his friend, John Cleaver, behaves with a superior air and keeps his distance from the locals. Cleaver's scepticism is eventually undone by the affection of a young mountain girl, and he is heartbroken when the girl tragically dies from an illness. In her innocence and simplicity, the girl serves as the very embodiment of the Southern Appalachians, and after having returned to the city, in his mind Cleaver keeps returning to the mountains. Ultimately, his experiences in the wilderness and the simplicity and sincerity of mountain living have broken the illusions of cultural superiority that he first entertained.

But Murfree's stories did more than merely portray the landscapes of the Smoky Mountains in the romantic tradition; they established a cultural contact zone in which mountaineers and outsiders met, where nature and culture, as well as past and future converged. Cleaver and Trelawney, for example, bring medical science to the mountaineers. Likewise, in her 1894 novel *His Vanished Star* a whole mountain community, human and animal alike, become involved in an outsider's plan to build a tourist resort on one of the mountain tops. Although Kenniston's hotel burns down by accident early in the novel—the work of a curious bear—and is never rebuilt, the construction process itself is described through an interesting merger of workers, visiting locals, and animals showing up at the building site:

> [From] far and near the mountaineers visited the unfinished structure. Often a wagon with a yoke of oxen would stand … while the jeans-clad owner would patrol the new building, solemnly stepping from timber to timber over the depths of the cellar, or with utmost simplicity of assurance make a critical circuit about the whole, and offer suggestions looking toward improvements. Sometimes the visitor was of shyer gentry: a red fox was glimpsed early one morning, with brush in air,

> speeding along the joists of the ball-room; it might seem they would never know the weight of aught more graceful or agile; a deer, doubtless a familiar of the springs, was visible once, leaping wildly down the rocks in great elastic bounds, evidently hitherto unaware of the invaders of these pre-empted sylvan wilds. Others, too, of the ancient owners of the soil came on more prosaic quest, but in the dead hours of darkness or the light of the midnight moon. A young bear … was brought to a pause, in a cautious reconnoitre, by the fragments of food, scraps from the workmen's dinner, which might be found by nosing about among the shavings. (166-167)

Although the fire that consumes the hotel is caused by the bear and might suggest a reading of the scene as nature's refusal of Kenniston's building, Murfree seems to accept the project without judgement, calling attention instead to the merger of interests, economic and environmental, invested in the construction.

His Vanished Star also provides one of the first Appalachian examples of the shift from an unintended pastoral to a designed pastoralized landscape. One of the key issues in the novel is Kenniston's concern that the small cottage of Luther Tem, an old mountaineer whose land adjoins his own, will spoil the scenery that he is trying to sell his future guests. Consequently, Kenniston spends much time and energy in trying to convince the old mountaineer to sell. As critic and Appalachian scholar Elizabeth Engelhardt writes, with the entrepreneur Kenniston, Murfree created a character who is "fluent in the language of the tourist and voyeur," and who would recreate and rearrange "the landscape and the buildings so that visitors felt they had discovered abstracted, silent nature—beautiful, mysterious, and picturesque" (107-108). Through Kenniston's attempt to construct the kind of pastoral scenery that he believes the tourists will find appealing, the novel engages in a dialectic structure, negotiating between the unintended pastoral of Luther Tem's small cabin and the pastoralized space of Kenniston's hotel plans.

By the end of the nineteenth century, the industrial growth in the region heightened local awareness of environmental issues. Industrial smokestacks polluted the air, textile mills dumped effluents into the waterways, and mining and lumber companies wreaked havoc by indiscriminately cutting down forests and blasting away mountainsides. Such concerns were expressed in the literature of the time. In Grace MacGowan Cooke's novel *The Power and the Glory* from 1910, the ambiguities of this new merger of nature and culture are apparent. In it Johnnie Consadine, a poor mountain girl, seeks her independence and fortune in a suburb of the fictional city of Watauga by working in one of the new cotton mills. Tired of her family's reputation for borrowing Johnnie wants to better her family's social standing. As Johnnie looks down on the city from afar, her vision is one of an alluring merger of nature and cityscape:

> It was Johnnie's first view of a big valley, a river, or a city. She had seen the shoe-string creek bottoms between the endless mountains among which she was born and bred ... she had been taken to Bledsoe, the tiny settlement at the foot of the Unaka Old Bald, where there were two stores, a blacksmith shop, the post-office and the church. Below her, now beginning to glow in the evening light, opened out one of the finest valleys of the southern Appalachees. Lapped in it, far off, shrouded with rosy mist which she did not identify as transmuted coal smoke, a city lay, fretted with spires, already sparkling with electric lights, set like a glittering boss of jewels in the broad curve of a shining river. (25-26)

From a distance the scene seems to reveal a promising middle landscape that suggests an ideal balance between nature and industry. As she soon learns, however, that balance is largely illusory. Her distance from the scene along with her ignorance of such matters means that she cannot identify the smog enveloping the city for what it is. Soon her experiences in the mill and in the worker's ghetto problematize her vision of the city. As Cooke shows, the new factory jobs were a mixed blessing. Industrialization helped break the isolation of the mountains and possibly even the bonds of social heritage, but also involved exploitation of workers and environmental destruction.

Nevertheless, the idealized blend of technology and nature contains possibilities to which neither Johnnie nor the author is blind. At the end of the novel, for example, Johnnie gets rich when her uncle Pros finds a long lost nickel mine, and the protagonist is also infatuated with the new automobile culture that was developing. Of course, at the time *The Power and the Glory* was published America had not yet witnessed the environmental problems unleashed by that particular technology, and Johnnie's attraction to the new technology serves both as a testimony to the increased accessibility to nature that the automobile provided at the turn of the century, and as a hint of a more general technological optimism.

Somewhat more sceptical about the wonders of technology was the former Saint Louis librarian Horace Kephart, who in time would become the most famous chronicler of Appalachian life and culture. Prior to his arrival in East Tennessee, Kephart lived a quiet life with his wife and children. After losing his job due to excessive drinking and because he occasionally disappeared to roam the nearby woods, Kephart looked to restore his physical and mental health, and to create a literary career for himself, by going to live in the wilderness. He decided to pick out the emptiest spot on the map and travel there, and since almost nothing had been written about the Southern Appalachians—a region almost the size of New England—that became his destination.

Nature was less benevolent and healing than Kephart had imagined, and upon arriving at his new mountain home in 1904, he was so exhausted that he had to be nursed back to health by a local mountaineer. Having overcome his fatigue he slowly began making a name for himself (Brown, *The Wild East* 82). Travelling the region, Kephart wrote articles about the great outdoors, and even a couple of wilderness manuals for sportsmen and campers, as well as a well-known collection of essays and stories about life in the mountains called *Our Southern Highlanders*. Educating the public on hunting, fishing and camping, Kephart shared his enthusiasm for the wilderness with his readers, and furthered the influx of tourists into the valley. Although it seems unlikely that the hard and often lonely mountain life could have done much, if anything, to improve his physical health, the masculinist narratives of bear hunts, the moonshining industry, and chasing escaped convicts, left his readers with no sense of his weak health or his drinking problems. Thus, at least through his work Kephart experienced the rebirth he had hoped for in the wilderness.

In keeping with his somewhat romantic views on nature, Kephart saw the industrial encroachment on his Edenic utopia as an evil that had to be stopped. After witnessing a train slowly making its way across the mountains, Kephart conveyed the horror of the experience to his readers in biblical terms. "Away down in the rear I heard the snort of a locomotive, one of those cog-wheel affairs that are specially built for mountain climbing. With a steam-loader and three camps of a hundred men each, it was despoiling the Tennessee forest. Slowly, but inexorably, a leviathan was crawling into the wilderness and was soon to consume it" (Kephart 104). By the early twentieth century an increasing range of industries had begun locating by the banks of the Tennessee River and its tributaries, and the natural resources of the mountains and the forests became increasingly accessible to lumber and mining companies. In 1910 Kephart even suffered the grim fate of having to leave his home on Hazel Creek when a lumber company purchased the lands and began operations.

Industrialization provoked concerned citizens like Kephart to ponder the possibility of preserving the Smoky Mountain wilderness for future generations. The idea for a national park located in the Southern Appalachians had first been proposed as early as 1885, albeit in another form, and it was seriously considered by the North Carolina state legislature in the 1890s, but without results (Pierce, *The Great Smokies* 37-38). It was not until the creation of the National Park Service (NPS) in 1916 that an official political framework for national parks took shape, yet without political backing the NPS could do little. Southern congressmen were traditionally sceptical when it came to federal matters, and the conservation-type projects that were typically suggested by progressive politicians of the time clashed with issues of state sovereignty and property rights. Nevertheless, the new director of the NPS, Stephen Mather, hoped that by lobbying the Southern politicians on behalf of a large national park project located in the South, they might be convinced to support other ventures as well (Pierce, *The Great Smokies* 48-50).

Ever since his appointment as director of NPS, Mather had campaigned to revitalize the national park movement by portraying the parks in a romantic light. Standing to gain from the transportation of tourists to and from the scenic wonders in the parks, the western railroads supported the campaign, and some even launched their own public relations campaigns. In 1906, the Great Northern Railway, for example, promoted landscape tourism with the slogan "See America First." Still, without broad political and local support, Mather realized, the task of creating a national park in the South was almost impossible; and indeed very little happened until the early 1920s.

Then in 1923 Anne and Willis P. Davis of Knoxville brought the idea of a Smoky Mountains national park back into the public debate. The couple had returned from a visit to Yellowstone National Park where they had noticed the booming trade that was going on around the park. The couple also wanted to protect the eastern mountains against the increasing industrial exploitation, but their main focus was the economic benefits associated with such large-scale tourism. To them, the Appalachians were just as beautiful as the mountains in the West and there was no reason that a national park in the East should not be as spectacular, and subsequently as profitable, as the ones already established in the West.

It did not take the Davises long to attract support for the creation of a national park in the Smoky Mountains. Willis was on the board of directors of both the Knoxville Chamber of Commerce and of the Knoxville Automobile Club, whose members soon recognized that a national park meant new and better roads. Anne was active in the League of Women Voters and the Knoxville Garden Club and in 1924 even gained election as a Republican member of the House of Representatives of the Tennessee State Legislature, where she sponsored a bill authorizing the purchase of 78,131 acres of mountain land from a local lumber company (Pierce, "Anne M. Davis"). It did not take long to rally local businesses and civic organizations, most of which quickly realized the tourism potential of the idea. A booster organization called the Smoky Mountains Conservation Association (later the Great Smoky Mountains Conservation Association, or GSMCA) was established, and a New York publicity agency was hired to promote the park and solicit donations. As historian Margaret Lynn Brown suggests, it might even have been the publicity agency that first inserted the word "Great" in front of the park's name (88).

Although the task of convincing locals to rally in favor of the park turned out to be relatively easy, people outside of those areas directly affected by the project were less willing to support it. Not only did promoters have to fight opponents backed by powerful lumber and paper companies, but the states of Tennessee, North Carolina, Georgia, and Virginia also competed over the location of the park. The states were even internally divided. Western Tennesseans, for example, saw no point in supporting something that would mainly benefit the eastern part of the state. And it was not until a couple of years had passed that Knoxville and Asheville boosters finally agreed to locate the park between the two cities in the heart of the Smokies.

In order to attract publicity, in 1925 Kephart was elected field secretary for the Asheville based booster organization the Great Smoky Mountains, Inc. He was a highly respected writer, and newspapers from both Knoxville and Asheville had already published several articles by him about the importance of preserving the Smoky Mountains wilderness. As someone who knew and loved the mountains, and who had witnessed their destruction personally, he was considered an authority on the subject. By early 1926 the many hours spent campaigning, lobbying, and travelling the countryside to promote the project had shifted the public and political opinion in its favor. A group of almost 170 Knoxville park boosters, including one high school band with a Smoky Mountains eagle as mascot, travelled as far as Florida to promote the cause (Pierce, *The Great Smokies* 103). Early fundraising efforts were also successful in collecting the amount needed to convince legislators of the seriousness of the proposal, and on May 22 of 1926 President Coolidge signed the bill calling for the creation of a national park in the Smoky Mountains. The bill allowed for the NPS to assume responsibility for the administration, development, and protection of the park as soon as 150,000 acres had been purchased. It was still up to local boosters to raise the funds necessary to purchase the remaining land, but when, in 1928, John D. Rockefeller Jr.—a man known for his support of conservation and philanthropic projects—offered to pay half of the ten million dollars the park was estimated to cost, the Great Smoky Mountains National Park was well on its way to becoming reality. However, Kephart was killed in a car crash in 1931 before he had a chance to see the park finished.

As opposed to the national parks in the West which were mostly located on government or state-owned lands, by the early twentieth century large parts of the Smoky Mountains were home to farmers and different industries, including lumber companies like the Little River Lumber Company which, according to Daniel Pierce, "retained timber rights to almost 16,000 acres of the 76,000 acres it had sold to the State of Tennessee in 1927" and continued logging operations within the boundaries of the planned park area late into the 1930s (*The Great Smokies* 134). Much land was also owned by members of the

Appalachian Club and the Wonderland Club, two locally based resort clubs whose members had built summer homes in the Southern Appalachians in the early part of the twentieth century (Brown, *The Wild East* 87). To claim the land needed for the park, thousands of people living and working in the area would have to be moved.

Consequently, in many places the "wilderness" which the park was to simultaneously showcase and protect would have to be recreated first. In the Smoky Mountains the shaping of visitors' experiences opened up for different interpretations of how the area could be formed. Early plans for the new park clearly reflected the economic emphasis that developers placed on it. It was necessary to reforest the region's woodlands and restock the local fish and wildlife populations, but to take full advantage of the tourism potential many park advocates desired a type of pastoralized middle landscape that would appear wild but which would also appeal to modern families. Golf courses, hotels, and landing strips for small air-planes within the park were some of the suggestions contemplated, as was flooding scenic areas in order to create the artificial lakes that attracted anglers and other sportsmen. Such plans, it seemed, were only defeated because it was not possible to find sufficient funding (Brown, *The Wild East* 120).

The lack of funding for specific park projects naturally posed a problem to park developers, and somewhat ironically it was a problem which the circumstances of the Great Depression helped solve. In 1933 Roosevelt created the Civilian Conservation Corps (CCC), putting young men to work on federal projects. Enrolling in the "tree army," as the CCC became known, gave participants a much needed pay check and helped them develop skills through job training. Between 1933 and 1942 no fewer than three million young men between the ages of 17 and 23 were assigned to a host of different federal projects. One of the first CCC assignments was the Great Smoky Mountains National Park, and by 1934 it housed 4350 workers in 17 camps. The CCC did allow African Americans to join, but no camps were created for black workers in the South since community leaders warned park officials that it might incite trouble in the otherwise segregated region (Jolley 9). This also reinforced the historical fantasy that the mountain South had always been racially pure.

Many different jobs awaited the CCC workers. Dead trees were removed from the forests and turned into firewood and building materials, and new trees were planted. They built fire-towers, telephone lines, bridges, and hundreds of miles of trail, sometimes painstakingly hand-carved out of bare rock. By the late 1930s 800 miles of trail and four major road systems had been constructed with CCC assistance.

The work was orchestrated according to a larger design. Three landscape architects, three junior landscape architects, eleven landscape technicians, and three planting foremen planned the construction of "trail shelters, campgrounds, amphitheaters, comfort stations, observation towers, and picnic areas" to ensure that the area appeared as natural as possible (Brown, *The Wild East* 127). They only used materials indigenous to the area in the construction of support structures like shelters and fences. Everything had to be "handcrafted from local rough-hewn materials with lines reminiscent of, if overscaled from, pioneer … construction techniques of the region," and the park headquarters were constructed in the style of "early-nineteenth-century eastern Tennessee stone houses" (Brown, *The Wild East* 131).

To the NPS, visions of the region's history became central to the new park's appeal. Pioneer buildings were restored and displayed within the park, adding a sense of history and human interest to the new middle landscape. Places like the Oconaluftee Pioneer Village or the restored Cable Mill helped convey a sense of cultural legacy to the tourists, even if it was a legacy cleansed of anything slightly modern: barbed wire fences, for example, were replaced by split-rail fences because the latter had a more rustic feel. When creating the popular Cades Cove section of the park, modern homes, farms, and stores were torn down to make way for carefully restored "primitive log cabins, barns, and churches that captured the pioneer image" the NPS wanted to sell (Martin 201). As historian Durwood Dunn has shown, although Cades Cove inhabitants—in the period between 1923 and 1927—had continuously been promised that they would be allowed to remain within the boundaries of the park, in the end the small community did not fit the vision that the National Park Service wanted to sell. In early 1936 the last remaining residents were evicted, but by then many of the former fields had already reverted to a wilderness condition. Park officials soon understood that it was precisely the contrast between the cultivated lands and the wild mountains that gave Cades Cove its unique charm. As a result, two studies were conducted on how to preserve the history of the region. It was decided that all buildings of a newer date that did not fit into the historicized narrative should be torn down, while older structures were allowed to remain as signifiers of a pioneer life frozen in time (Dunn 243-257). The overall result of the Great Smoky Mountains National Park was an often staged and idealized image of a past clothed in an appealing cultural simplicity. As Margaret Lynn Brown notes, park officials created "a scene, a rustic image, a vision of what America should be. To accomplish this, they revised and adulterated history in a way that suited their own interests and tourists' aesthetics" (138).

The recreation of wilderness in the Great Smoky Mountains National Park was not a new proposition. Similar design issues had been part of the preservation efforts in Yellowstone, Yosemite, and Glacier National Park, where efforts had often required the relocation of Native Americans living within the park boundaries.[1] In Yosemite, as in the Smoky Mountains, nature had to be recreated before it could be properly enjoyed by tourists. Since its discovery by white Americans in the late 1840s the exploitation of the area's resources by local farmers and entrepreneurs had left parts of the landscape in a less attractive state. Free-grazing livestock had destroyed the wildflowers, fences and barns defiled otherwise breathtaking views, and one hotel owner had even "cut a swath through the trees to provide his barroom with an unobstructed view of Yosemite Falls" (Runte 59). As early as 1865 Frederick Law Olmsted, having chaired a commission to recommend plans for the future use and upkeep of the Yosemite area, presented a report where he, among other things, noted that the markings made by visitors or native peoples on rocks and trees threatened to rob the area of its pristine beauty (Olmsted). To prevent such destruction while opening the area up to future visitors, Olmsted suggested the construction of a public road, trails hidden from sight by trees so as not to interrupt the panoramic views, as well as bridges and lookout points to help "shape visitors' experience of Yosemite by directing their movement and gaze" (Spirn 93).

[1] For a detailed history of the removal of Native Americans from national park lands, see Mark David Spence, *Dispossessing the Wilderness: Indian Removal and the Making of National Parks*

The combined desire to simultaneously reclaim the Smoky Mountains wilderness by reforestation efforts and fish and game programs, while reaping the economic benefits of the tourist industry, resulted in the creation of a middle landscape where both history and the future were on display. Not only did the park showcase old pioneer style dwellings and technology, it also featured new and modern roads with strategically located viewpoints, where tourists could experience the Smoky Mountains majesty from the comfort of their cars. But while the Great Smoky Mountains National Park was among the first large-scale projects to pastoralize the Tennessee Valley, it was not the only one. From the 1890s until the early 1930s the region also saw the creation of George Washington Vanderbilt's impressive Biltmore Estate, as well as Benton MacKaye's proposed Appalachian Trail, both of which were based on pastoral simulations.

II

There is plenty of scenery to catch any tourist's attention along Interstate Highway 40 as it weaves through the Smoky Mountains towards Asheville, North Carolina. The route not only allows for a closer look at the hauntingly beautiful and wild character of the Smoky Mountains. If the goal of the trip is the well-groomed grounds of the Biltmore Estate, the drive also makes for an interesting transition from the wilder appearance of the Pisgah National Forest to the gentler pastoral scenery at the Vanderbilt estate. Both places, however, are part of the middle landscape tradition that people like George Washington Vanderbilt, Frederick Law Olmsted, Richard Morris Hunt, Gifford Pinchot, and Carl Alwin Schenck helped introduce to the Southern Appalachians in the late nineteenth and early twentieth century.

Biltmore is located at a general altitude of over 2000 feet, in the midst of mountains and dark forests, and the estate epitomizes the very essence of the pastoral experience, constituting what Char Miller has called "a large-scale aesthetic experiment intended to reconcile nature and culture, the wild and the civilized" (Miller, *Gifford Pinchot* 102). Surrounding Biltmore lies an open and green landscape designed, beautified, and managed in order to fulfil every Edenic ideal. A tour of "America's largest home" includes not only the

French Renaissance style mansion with its 250 rooms and four acres of floor space, but also a wooded park, five pleasure gardens, several working farms, thousands of acres of scientifically run forest lands, and a winery, which was added to the estate as recently as 1985.

Biltmore was the result of millionaire George Washington Vanderbilt's fascination with the extravagance of the European country estates that he had visited on his many travels. Born into one of the nation's wealthiest families, George Washington Vanderbilt's fortune was primarily a result of the enterprising skills of his grandfather, Cornelius Vanderbilt, more commonly known as the Commodore. As a young boy in early nineteenth century New York, Cornelius Vanderbilt made the family's first fortune when he turned a small New York ferry service into a fleet of more than a hundred steamboats that sailed as far as Europe and South America. Later in life the Commodore made his second fortune by investing in the growing railroad industry, and at the time of his death in 1877 he left his only living son, William, more than 100 million dollars and an industrial empire.

Fig. 4.: Biltmore Estate, 1902. William Henry Jackson, photographer. Courtesy of the ***Library of Congress Prints and Photographs Division, Washington, D.C. 20540 USA.***

George Washington Vanderbilt, the youngest of the Commodore's grandchildren, was wholly uninterested in the family business, and instead spent much of his time travelling the world. As a learned man, and somewhat inspired by his family's tradition of large and impressive homes, Vanderbilt envisioned a magnificent home of his own that could function both as a showcase for the many collections he had gathered during his travels to Europe, Africa, and Asia, and at the same time serve as a social center for his family and friends. In the late 1880s, during one of his travels, he fell in love with the mountain vistas and the benevolent climate of western North Carolina, and since land in the region was cheap, Vanderbilt saw the perfect opportunity to fulfil his dream there (*A Guide to the Biltmore Estate* 12-13).

In the following years Vanderbilt began purchasing land in the area, which also meant buying out a number of impoverished planters who resided in the forested mountains. Starting with a mere 2000 acres of land, at its height Biltmore comprised well over 100,000 acres of mountain land that, apart from the estate itself, included approximately 7000 acres of Biltmore Forest, and an additional 80,000 acre tract which would later become the Pisgah National Forest.

Only one year after his visit, construction on the estate had begun. Vanderbilt hired two of the best known architects for the job. Richard Morris Hunt was legendary as the founder of the first American school for young architects and had won praise for the Lenox Library in New York as well as for designing the base for the Statue of Liberty. He was also among the organizers of the American Institute of Architecture, of which he became president in 1888. He had studied at the École des Beaux-Arts in Paris and was stylistically drawn to the Beaux-Arts movement and its richly ornamented, symmetrical, and classical designs. Frederick Law Olmsted, who was hired to design the landscapes surrounding the mansion, had won national acclaim for his designs of urban recreational spaces, including Central Park in New York, and by the late 1880s was considered one the most influential landscape architects in America. He had no formal training in either gardening or architecture, but he had an eye for landscapes and early in life had made an impression upon parts of the Staten Island gentry with the work he had done at his Tosomock Farm; it was also at this time that he became friends with William Vanderbilt. Having always loved the outdoors, on his travels abroad Olmsted was an eager student of European gardening. The

English landscapes fascinated him in particular, and his visit to the Birkenhead Park near Liverpool in 1850 proved a major source of inspiration. There, on what had once been a flat and useless piece of land, the English architect Joseph Paxton had created "picturesque ponds, random clumps of trees, rolling meadows, overgrown hillocks, and meandering footpaths [that] reminded Olmsted of the English countryside" (Rybczynski 93).

After returning home, Olmsted spent a significant part of the 1850s travelling the United States, writing books and articles which commented on everything from nature to political matters such as slavery. But in 1858 he was to get his chance. Olmsted and an acquaintance, Calvert Vaux, entered a design contest for what would become Central Park in New York, and won. It was an important first step towards earning Olmsted his future reputation, because it proved that not only did he have an understanding for the complexity of landscape design, he was also able to transform such ideas into a working plan. In the following decades the jobs he was offered included such prominent places as the Niagara Reservation at Niagara Falls, the grounds of the Capitol in Washington D.C., and Boston's park system. Not long after he was employed at Biltmore, Olmsted was also commissioned to design the grounds for the 1893 World's Fair in Chicago, a dual engagement that kept the almost seventy-year old landscape architect travelling between Chicago and North Carolina.

Olmsted's involvement at Biltmore began in August of 1888 when Vanderbilt invited him on a trip in order to inspect his newly purchased lands in western North Carolina, and comment on his plans to build a large pastoral estate there. Olmsted had to disappoint the young millionaire, by informing him that although the air and the view were excellent, in many places the quality of the soil and the tree crops were too poor, and the topography was unsuitable for establishing a park (Rybczynski 380). Parts of the property were simply too rugged for Olmsted to work with, and in other places extensive overfarming along with the practice of using controlled fires to remove underbrush and tree stumps had ruined soil and trees. Instead he advised Vanderbilt to set aside a portion of the grounds for a forestry practice that could potentially set an example for the rest of the nation to follow. Olmsted had long been an advocate of forestry in America, and in 1871 had written an article on the relationship between forests and water supply. He was also close friends with the one of nation's leading dendrologists, Charles Sprague Sargent, and consequently understood the potential for forestry (Roper 415).

At Biltmore Olmsted created a varied park landscape in the style of the English pastoral tradition. His design focused on a series of diverse ornamental gardens including an Italian Garden, a Shrub Garden, a Walled Garden, an Azalea Garden, and a Spring Garden. He also created a large open landscape with lakes and trees known as the Deer Park, where Vanderbilt and guests could engage in horseback riding or even hunting, and he constructed four walking trails to make sure that the estate could be enjoyed to its fullest. Olmsted also made plans for a couple of working farms and the Biltmore Arboretum where flowers and plants from all over the world were to be displayed along with a large collection of different species of trees that were tested for scenic and forestry-related qualities.

One likely reason for Biltmore's popularity may well be the sense of history that Vanderbilt literally had built into the walls and gardens of the estate. Olmsted carefully sought to give the surrounding forests an ancient look, just as Richard Morris Hunt sought to make the estate appear older by using wood panelling (Miller, *Gifford Pinchot* 102). The style of the extravagant mansion imitated that of the landed European aristocracy that Vanderbilt so often visited on his travels, and it allowed him to feel like an established part of that cultural tradition. In a sense, Vanderbilt sought to inscribe Biltmore into a context that could be called both undemocratic and un-American. The estate evoked the hierarchies and decadence of Europe rather than the equalizing and democratic force with which the American landscape has traditionally been associated.

Olmsted also spent much energy on designing the three mile long Approach Road that would lead visitors to the mansion itself. Entering the grounds through the Lodge Gate, the visitor would find himself on a long scenic road designed to shield the mansion from sight until the last possible moment, creating a dramatic transition between forest and the sudden openness of the rectangular front lawn and the mountain scenery in the back. As the official guide to Biltmore describes it, the road "snakes along the ravines through dense border plantings of rhododendron, mountain laurel, and azalea, passing from woods to open meadow and back again to groves of hemlock and pine—with no distant views to interrupt the intimate effect. At every turn is a new surprise: a stream, a pool, a blanket of wildflowers, a thicket of river cane" (*A Guide to the Biltmore Estate* 83). As with the rest of the Biltmore landscapes, the entrance road was designed to blend into a natural setting, creating the appearance of an untouched environment. As Anne Whiston Spirn notes, driving along "the entrance road through a lush, mature forest, one finds it difficult to imagine that this landscape was constructed" (99).

Fig. 5: Biltmore approach road, 1895.
Courtesy of the Frances Loeb Library, Graduate School of Design, Harvard University.

The new estate immediately caught the attention of the local communities. Many were naturally sceptical about the ideas of the northern millionaire, and most were even convinced that Vanderbilt could never make his plan work and that the attempt would bankrupt him. The terrain simply was not suited for the kind of farming and gardening that were proposed at Biltmore. But as work progressed views regarding the project began to change, and as one newspaper article noted, by the late 1890s Vanderbilt had come to be regarded by many as "the great benefactor of Western North Carolina … [who] has shown the Carolinians the productive capacities of their Virgin Soil (that great mine of wealth they have profited by so little) by the scientific drainage, the improved machinery, the importation of fine stock, the judicious and lavish use of fertilizers, and the most up-to-date and scientific methods of farming that have characterized the management of the farm from the beginning." The article went on to emphasize the stunning transformation of the wretched wilderness that characterized the region before Vanderbilt arrived. "Anyone who knew the barren hills, the washed rut gullies, sedgefields, swamps, ditches, and the succession of worn out farms with their tumble-down houses—in which their owners were starving—that a short while occupied the site of the now splendid and fertile Biltmore Estate is struck with amazement at the marvellous change that has been wrought by the wise expenditure of money on the most desirable and beautiful location in the world" ("Farmer Vanderbilt" 1). In the context of Vanderbilt's attempts to have his estate imbued with a sense of history, the article's emphasis on the ahistorical nature of Biltmore is interesting. In truth, the realities of a rugged and infertile land, under which many of the area's former residents had suffered, clashed with the designed fertility of the new estate's parks and gardens. It was only through science and vast amounts of cash that nature was changed into the pastoralized landscape that tourists can still marvel at today.

But not all of Biltmore was about gardens and pastures. Vanderbilt took Olmsted's idea of creating a forestry practice to heart, and in 1891 he hired the young American forester Gifford Pinchot to revitalize the forest areas and work out a plan for the management of Biltmore Forest and the surrounding Pisgah Forest. Pinchot was born in Connecticut in 1865 into a family so wealthy that he had no need for a job, yet the idea of merely enjoying his family's fortune never appealed to him. Upon his graduation from Yale in 1889, Pinchot went abroad to further study the subject of forestry in Europe. The desire to study that topic at a time when no American university offered such courses came from his father who firmly believed in the idea. James Pinchot was a friend of both Richard Morris Hunt and Olmsted, just as he was a great admirer of the Hudson River School paintings of which he had a collection. Few people in nineteenth century America saw forestry as having much of a future in a country where forests, along with other natural resources, were perceived as abundant, but with his father's support, Pinchot was adamant about his career choice and soon travelled to Europe to learn more about forestry.

Pinchot's original plan for going to Europe was merely to collect books and writings on forestry, and to visit with such renowned names in European forestry as Sir William Schlich, head of the Forest School at Cooper's Hill in England, and the German forester Sir Dietrich Brandis. But during his travels Pinchot was unexpectedly offered the chance to enroll as a foreign student in the Ecole Nationale Forestière in Nancy. Excited about the opportunity to study at the French forest school and already fluent in the French language, Pinchot quickly agreed to spend the following year in France, attending classes and visiting different European forests. Pinchot's stay in Nancy taught him much about the scientific management of forests, including how to produce the best possible tree crops and how to make the forest pay. But socially the ambitious American had only scorn for the careless manner of the French students who seemed not to care about their chosen profession, and as a result Pinchot spent much of his time walking the forests, linking the theories he had learnt to his own observations (Miller, *Gifford Pinchot* 78-84).

After a year in France, Pinchot was ready to return home in spite of his professors' warnings that he had not nearly enough formal training to manage a forest. But Pinchot was undeterred and upon reaching America started working a few minor jobs that he felt might afford him more experience. As suggested by his father he began work on a book about forestry, but he also gained employment from Phelps Dodge & Company, who hired him to write a report on the possible introduction of forestry practices on some of the company's lands in Arizona. Given a chance to travel the country, he used the opportunity to visit Biltmore Estate. Even though it was still under construction when Pinchot arrived, Biltmore nevertheless was a magnificent sight. No expense had been spared; the castle was built entirely out of limestone that had been brought in from Indiana, more than 600 miles away. Although Pinchot later wrote that the placement of such an overwhelming estate in a region otherwise characterized by small dirt farms and log cabins served as "a devastating commentary on the injustice of concentrated wealth," upon first visiting the estate he excitedly noted in his diary that it was "just right for forest management on a rather intensive plan" (Pinchot, *Breaking New Ground* 48; Pinchot, *The Conservation Diaries* 51).

During 1891 Pinchot had several meetings with Vanderbilt and Olmsted about introducing forestry on the Biltmore grounds. Pinchot was thrilled about the opportunity. Here was a chance to implement his ideas. As he recalled in his memoirs, *Breaking New Ground*, "Biltmore could be made to prove what America did not yet understand, that trees could be cut and the forest preserved at one and the same time" (49). Pinchot's plans for the estate made an impression upon Vanderbilt and Olmsted, and later that year he was hired.

But managing Vanderbilt's forests was slow work, as Pinchot makes clear in a booklet he produced for the Columbian Exposition in Chicago in 1893. The booklet was to explain to the public the forestry work that had been done at Biltmore during the first year. Pinchot's approach to creating the pamphlet also shows an aesthetic side to his understanding of nature. A little known fact about Pinchot was his interest in photography, but as Char Miller explains, Pinchot often took pictures during his travels. Part of his strategy, not just in Chicago but in general, was to convince people that Forestry was needed to conserve natural resources and scenic landscapes for future generations by showing them "images of badly burned and grazed lands, soaring forests of hemlock or fir or pine, and stunning panoramas" (Miller, "Char Miller on Gifford Pinchot").

But the booklet also gives an interesting insight into the problems with which the young forester was dealing. Much of the land was exhausted due to earlier farming practices, and the trees in Biltmore Forest were generally of poor quality. As a result, Pinchot first established a series of improvement cuttings in order to even out the forest. His goal was a forest in which the different age groups grew together, but in many places older growths mixed with younger trees in ways that impaired their growth and made it difficult to work out a plan for the whole forest (Pinchot, *Biltmore Forest* 18). Furthermore, Pinchot excluded cattle from the forest, since they damaged the young trees, forbade the use of fire, and established areas that needed rest in order to be productive again.

Economically, Biltmore Forest was not as big a success as Vanderbilt had probably hoped, but Pinchot did not despair. As he pointed out, much of the wood available in Biltmore Forest when he took over was only good for firewood, and since the area in which the estate was located had plenty of firewood, it was difficult to sell. Consequently, the first year's profits on the forest were a mere $392.40, and that was only possible because Pinchot included the value of the stock on hand that had not yet been sold (Pinchot, *Biltmore Forest* 39-40).

The slow progress at Biltmore proved increasingly frustrating for the young forester. But his arrangement with Vanderbilt allowed him to also work elsewhere, and in 1893 Pinchot opened an office in New York offering his services as a consulting forester. Another compelling reason for leaving his position at Biltmore was the untimely death of his wife Laura Houghteling who had suffered from tuberculosis. Laura spent her last months in her family's house by the French Broad River not far from Biltmore, and although Pinchot did not seem to openly grieve Laura's death, it must have been a devastating experience for the young man. Indeed, he spent the next twenty years of his life being faithful to his dead wife, and often wrote about her in his diary as if she was still alive (Bradley 207-208).

To replace Pinchot, Sir Dietrich Brandis recommended a young German forester by the name of Carl Alwin Schenck. Initially Pinchot admitted to being pleased with the choice of Schenck, and although he felt that the young German was rather rash in some of his conclusions, he also believed that Schenck was improving (Pinchot, *The Conservation Diaries* 60). But as time went by the relationship between the two men deteriorated. Pinchot came to believe that Schenck's background in German forestry had scarcely prepared him for the challenges of an American forest, and Schenck felt that Pinchot seemed oddly uninterested in the work at Biltmore, talking more about recreational activities such as hunting and fishing than about the problems at hand (Schenck 28). The professional relationship between the two foresters was also strained. Initially Pinchot had explained that he would remain in the position as Chief Forester and that Schenck was to work under his supervision (Schenck 20-21). Vanderbilt, however, informed Schenck that Pinchot's relation to Biltmore had been terminated, and that Schenck was free to manage the estate's forest according to his own designs (Schenck 34).

Fig. 6: Gifford Pinchot, 1890—1910. Courtesy of the Frances Benjamin Johnston Collection at the Library of Congress Prints and Photographs Division, Washington, D.C. 20540 USA.

In many ways Schenck was unimpressed by the progress that Pinchot had made at Biltmore, most of which he summed up as the mere removal of "the large number of dead and dying trees … found all over the estate" (25). To the disciplined German there seemed to be little organization in what Pinchot had done, and there was no established forest nursery in which he could grow a planting stock, no roads had been built on which to transport the lumber, and many of the trees were of such poor quality that they were simply unfit for the market. A major point of contention was the construction of a splash dam for transporting logs down to the Mills River. To Schenck a splash dam was a temporary answer only and he felt that a road was not only a better solution, but that it would also solve the problem of transporting teams of loggers into the forest. But Pinchot's plans had already been initiated and Vanderbilt expressed a desire to see that project continued. Consequently, Schenck had the splash dam built, but it was a decision that would have dire consequences when, in 1897, the swollen rivers sent thousands of logs tearing into riverbanks and bridges, causing a number of local farmers to sue Vanderbilt for the damages to their property.

In spite of his problematic start, Schenck eventually won widespread recognition for his forest management. He installed experimental plantations of indigenous species, and constructed roads for lumber transportation. In 1898 he even founded the first forest school in America, the Biltmore Forest School, which offered students a one-year course and graduated over 400 students before it closed in 1913. Nevertheless, it was becoming increasingly clear that the Biltmore forestry experiment was not as profitable as hoped for. Vanderbilt's deteriorating financial situation following a bad investment in 1902, as well as the economic depression in 1907, led him to sell Pisgah Forest. The newly established Forest Service, the former Division of Forestry, of which Pinchot was the Chief, was interested in buying the land to establish a National Forest. Although the decision to sell Pisgah was not completed until after Vanderbilt's unexpected death following an operation for appendicitis in 1914, it marked the end of the Biltmore forestry experiment.

The merger of the pastoral ideal with a vision of a working conservation-based forestry enterprise at Biltmore epitomizes the pastoralized middle landscape in the late nineteenth and early twentieth century. Fascinated with the anti-modern, Biltmore's designs managed to integrate nostalgia with a progressive utopianism that recognized the need for resource planning. Although modelled, in part, after the European aristocratic tradition, in some ways Biltmore was not unlike the socialist inspired plans for an Appalachian Trail, which also sought to incorporate public planning with a pastoral vision.

III

The AT, as the Appalachian Trail is often called by those who hike it, is a 2174 mile long scenic mountain trail that stretches from Maine to Georgia. The trail is yearly walked by an estimated 3 to 4 million people, and to some constitutes "a spiritual domain, the setting for a rite of American cultural passage and pilgrimage" (Anderson, *Benton MacKaye* 5). In the foreword to a 1972 National Geographic publication about the trail, Benton MacKaye—the AT's father—considered the realization of the trail perhaps "unrivalled by any other single feat in the development of American outdoor recreation" (MacKaye, "Foreword" 5). As a testimony to the trail's significance in American culture, a quick search on the Internet reveals both documentary films about the trail, CDs with Appalachian Trail music (whatever that is), and a host of books ranging from guides on how to prepare for a hike, to scholarly analyses of the trail's history and meaning, popular accounts by people who have walked it, and even horror novels such as Stephen King's *The Girl Who Loved Tom Gordon*. The Appalachian Trail continues to engage the American imagination on several levels, even if it is far from the visionary and socialist inspired project that a young Benton MacKaye imagined in 1921.

As opposed to the Great Smoky Mountains National Park and the Biltmore Estate, the AT's relations to the Tennessee Valley are perhaps less tangible. Although Benton MacKaye would later work as a regional planner for the TVA, a project that combined his love for planning and conservation, he had no personal relation to the Tennessee Valley, and his ideas were mainly inspired by the New England landscapes that he knew and loved. Indeed, the trail's inclusion in this study mainly owes to the fact that approximately 400 miles of the trail wind through southern Virginia, along the Tennessee and North Carolina border, and into northern Georgia, crossing both the Great Smoky Mountains National Park and the TVA's Watauga Dam before reaching its southern terminus. Also, as opposed to some of the other places studied here, it is the vision of the pastoralized AT as a utopian middle landscape that is of interest, rather than the actual constructed trail.

Emile Benton Mackaye was born in Stamford, Connecticut in 1879. His father, Steele MacKaye, was a wilful, visionary, and talented playwright who had enjoyed some success on the stages in New York City, and had even helped set up Buffalo Bill's popular Wild West shows in Madison Square Garden in 1886 and 1887. In spite of his talents, Steele struggled to provide his family with financial security, and as a result the MacKayes were accustomed to moving about, living in New York and Washington D.C., as well as in more rural surroundings such as Stamford. Benton soon discovered that the urban milieu of New York held less interest for him than the natural environments that surrounded Shirley, Massachusetts, the small New England community in which he felt most at home. Inspired by legendary American explorers such as Major John Wesley Powell, whom he heard speak at the National Museum in Washington in 1891, he embarked on numerous expeditions in the vicinity of his home, and spent much of his time drawing maps of the area and writing about the landscapes he experienced. Given his love of the outdoors, it was hardly surprising when he eventually decided to study forestry, which the young Gifford Pinchot had succeeded in popularizing.

But before he got as far as receiving a formal training in forestry, years of college education awaited him. MacKaye only barely made it into Harvard, where he registered in October of 1896, and according to biographer Larry Anderson, in most respects he was an average and somewhat lazy student. But two introductory courses in geography and geology proved to MacKaye that it was possible to couple his love of nature with higher learning (Anderson, *Benton MacKaye* 29-34). MacKaye also began spending his vacations hiking in the New England mountains and countryside with a couple of friends from school. Apart from a few academic digressions into American history and politics, he continued to elect courses on earth science-related subjects. During these years he also began developing his sense of the American landscape as an open and democratic space, unsullied by factories and other urban developments.

MacKaye's hikes in the wilderness introduced him to sublime landscapes, but they also showed him the many threats to the eminent scenery, and the need to protect it from industrial exploitation. The demand for lumber in the post-Civil War economy meant that many states had begun selling forestland to privately owned lumber companies. This was a development that affected the Southern Appalachians as well as the woodlands of New England (Clark 14-16; Anderson, *Benton MacKaye* 35). The idea of opening up the landscape to people while protecting it from harm, coupled with his love for spending time in nature, pointed to a career in forestry, combining the aesthetic enjoyment of nature with commercial use through planning. But it was not until a near fatal case of appendicitis confronted MacKaye with his own mortality that he decided upon forestry as the career that would allow him to transform his life-long passion for nature into a profession (Anderson, *Benton MacKaye* 44-45).

By the time MacKaye finished at Harvard, Gifford Pinchot had left the Biltmore Estate and become head of the Division of Forestry. Pinchot's experiences at Biltmore had convinced him that public, rather than private, forestry was the way of the future, but he continued to travel the country, promoting the idea of scientifically managed tree crops. As a result, forestry was becoming an increasingly popular field, and whereas interested students had formerly had to travel to Europe to receive an education, in 1898 things began to change. Pinchot's successor at Biltmore, Carl Schenck had established his forestry school on the estate's grounds, and the same year Bernhard Fernow, Pinchot's predecessor as head of the Division of Forestry, started a four-year program at Cornell University. Pinchot still distrusted the German forestry tradition's potential for success in America, and he still desired to provide American foresters with a distinctly American approach to tree crop management, and so in 1900 Yale, with the financial backing of the Pinchot family, also started a graduate program in forestry (Pinchot, *Breaking New Ground* 152).

MacKaye had heard Pinchot speak about the scientific management of forests at Harvard in 1900, and he had felt instantly drawn to that profession. In 1902 Harvard also began offering forestry courses, and the next year MacKaye returned there to pursue such a career. By 1905 he had earned his master's degree, and the timing could not have been better for the young graduate. MacKaye finished just in time to benefit from the establishment of the Forest Service and the hundreds of new forestry-related jobs that were created as a result of the new Federal Forest Transfer Act, which placed the responsibility for the management of 63 million acres of national forestland under the new bureau.

MacKaye worked for the Forest Service until January 1918, when he was transferred to the Department of Labor, a job which he left again a year and a half later. He never made much of a name for himself in either the Forest Service or the Department of Labor, but he gained the respect of those who worked with him, and he was in a good position to develop his socialist inspired views on the interdependency of human communities and nature. In 1915 he had married women's rights activist Jessie Hardy Stubbs, whom he had met and fallen in love with the year before and who shared many of his political views.

MacKaye was surrounded professionally and socially by conservationists and Progressives, and for years he travelled the country surveying forest areas and talking to people who had been affected by the industrial exploitation of the land. It was during these years that he developed his approach to resource planning. As early as 1914 he had designed a plan for the Great Lakes region which called for "government-directed settlement, public ownership of land and resources, planning on the community level, and cooperative marketing" (Sutter 145). When he wrote the report "Employment and Natural Resources" for the Department of Labor in September of 1919, many of the core ideas that would later become central to his Appalachian Trail proposal could be found in there. MacKaye wanted to ensure the survival of local communities through government controlled resources and conservation practices. The depletion of natural resources by profit-driven privately owned companies would be replaced by a sustain-yield balance maintained by permanent forest, agricultural, or mining based communities.

Then in April of 1921 tragedy struck. His wife, suffering from depression, committed suicide by throwing herself in New York's East River. The loss took a great toll on Benton, and when he was invited to stay with a friend, Charles Harris Whitaker at his country retreat in New Jersey, he accepted the chance to get away for a while. Whitaker, who was an architect and the editor of the *Journal of the American Institute of Architects*, was also interested in planning and recreation, and while staying at the retreat MacKaye began working again. At the heart of his work was an idea for a trail spanning the entire length of the Appalachian Mountains. Whitaker encouraged the idea of an Appalachian Trail, and on July 10 he introduced MacKaye to his friend Clarence Stein, who was head of the American Institute of Architects' Committee on Community Planning (AIA) as well as the Society for Ethical Culture, a group that was engaged in social reforms addressing the problems created by industrial society.

The July 10 meeting accelerated the process. Whitaker promised to publish MacKaye's article about the trail in his journal, and Stein pledged to promote the idea in AIA circles. Three months later MacKaye's brief but visionary article "An Appalachian Trail: A Project in Regional Planning" appeared in the October issue of the

Journal of the American Institute of Architects. In the article MacKaye kept the original idea of an alternative to the prevailing capitalist and industrialist social structures intact, but he emphasized the recreational aspects of the project, to avoid attracting criticism from right-wing politicians and industry, a choice which undoubtedly also helped attract the support of many existing conservation groups.

MacKaye's trail plans were born out of the same post-frontier anxiety that haunted so many early twentieth-century American conservationists. He saw civilization as a weakening factor that rendered people increasingly helpless and disconnected, and believed that through re-establishing contact with nature it was possible to harvest "the strength of progress without its puniness" (MacKaye, "An Appalachian Trail" 371). In that sense, the pastoralized landscape that was the basis for the Appalachian Trail offered to solve the same problems that the Boy Scout movement, advocates of hunting, and various other outdoors enthusiasts had identified as America's crisis of modernity. MacKaye would later explain that the trail was meant as a way for modern Americans to become acquainted with the scenery and absorb the landscape as the "opposite of machine influence .... [and an] antidote for over-rapid mechanization" (MacKaye, "The Appalachian Trail" 330).

But MacKaye's project was more than a response to the disappearing wilderness. As opposed to many other nature enthusiasts at the time, MacKaye, like his mentor Pinchot, was a utilitarian who firmly believed that nature existed for the benefit of man, and that wise use was the best way to preserve natural resources. In many ways MacKaye's proposal constituted the ideal pastoralized landscape. He imagined a giant looking down upon the Appalachian Mountains and the Eastern seaboard, and what he saw was a landscape divided between undeveloped, or under-developed, lands and over-industrialized urban centers (MacKaye, "An Appalachian Trail 374). It was a landscape shaped by a belief that industry was a goal in itself rather than a means to an end, and it was this approach to nature, community, and life that MacKaye proposed to change by better integrating those two worlds. Instead of 25 million acres of undeveloped wilderness, MacKaye imagined "room for a whole new rural population. ... [and] an opportunity—if only the way can be found—for that counter migration from city to country that has so long been prayed for" (MacKaye, "An Appalachian Trail" 375).

The backbone of these new rural and recreational societies was a walking trail from Mount Washington in New Hampshire to Mount Mitchell in North Carolina. Along the trail a number of community camps would offer less strenuous activities for families, while food and farm camps would provide communally owned places where people could stay and volunteer to work for the good of the community. Transforming leisure time and recreation into work would not only allow participants a chance to escape the city, but it would allow people to experience the non-industrial life of a farmer, a miner, or a lumberjack. This, MacKaye hoped, could also revitalize the labor movement and help better balance the scales between industrial and non-industrial society (MacKaye, "An Appalachian Trail" 376). Last but not least, he figured that such a volunteer force potentially could prove the equivalent of 40,000 workers employed in continuous activity, a considerable and valuable asset to any society. MacKaye's was essentially a socialist vision with a pastoral slant.

The actual construction of the trail proved frustratingly slow. During the first year a few steps were taken towards completing individual parts of the northern trail. These initiatives were mostly undertaken by local hiking clubs inspired by the trail concept, but as a unified project little was achieved (Foresta 78). In 1923 MacKaye and Clarence Stein helped form a new group of social reformists and planners called the Regional Planning Association of America (RPAA). It served as a forum for new theories in planning, and it also functioned as a platform from which to seek financial and political support. Although counting upwards of 30 members, the RPAA was essentially a small group of friends that included MacKaye, Stein, and Lewis Mumford, who was already famous as an urban planner and a social philosopher. From this group came most of the intellectual and organizational ideas. The RPAA quickly adopted MacKaye's project as its first real initiative because the trail encompassed many of the characteristics that were seen as central to regional development.

Since the actual construction of the trail would have to be undertaken by many local hiking organizations, it was essential to create a group to oversee its construction and administration. As a result, in 1925 the Appalachian Trail Conference was formed. Structurally, the trail conference was a superstructure loosely made up of the local trail clubs and designed to handle communications with the federal government, as well as to standardize trail construction in the different states. The Appalachian Trail Conference was arranged on March 2, 1925 in Washington D.C., and it brought together a wide range of speakers. Some philosophized over the trail, others commented on its potential, and some merely described the progress of construction in their area. But the conference also elected the new trail administration. For unknown reasons MacKaye was not elected to be part of the governing body, which included such spirited characters as Myron Avery, Raymond H. Torrey, Horace Kephart, and Harvey Broome, all of whom had written about nature and recreation. However, common for many of those who became involved in the project, as Ronald Foresta notes, was that few suspected, or might have agreed with, the socialist ideology behind the trail. They were attracted to the idea of a sentimentalized past and the themes of anti-modernity and escapism (81-82). The trail had gone from being MacKaye's intellectual property to resting in the hands of a corps of volunteers who did not necessarily see eye to eye with his regional planning vision, but saw the trail as a means to briefly escape into nature.

It was, of course, ironic that the industrialism that caused so many of the problems that MacKaye's planning was directed at solving also was responsible for the social security that was needed to foster so many recreation-minded nature enthusiasts willing to spend their time and energy organizing and constructing the trail. The social ideology of Progressivism that had been such an integral part of the trail's ideology was being replaced by an individualistic approach to nature and recreation. To the new generation the development of an Appalachian Trail had little to do with social reform and everything to do with personal freedom; hiking was "an alternative to non-constructive urban leisure" (Foresta 84).

As work on the Appalachian Trail slowly progressed, MacKaye set about writing the book that had been on his mind for nearly a decade. *The New Exploration: A Philosophy of Regional Planning* was MacKaye's most ambitious work, calling for a replacement of old explorations that dealt with "actualities" with a new type of exploration that would concern itself with "potentialities" (30). The book also constituted a larger theoretical framework that fleshed out many ideas behind the trail concept. Although essentially a compendium of articles, manuscripts, and planning projects that MacKaye had completed during the 1920s, through Mumford's editorial suggestions it became an original and visionary work, that, according to the editor's introduction, deserved "a place on the same shelf that holds Henry Thoreau's *Walden* and George Perkins Marsh's *Man and Nature*."

In the book MacKaye promoted regional planning as more than an approach to the urban, the rural, and the undeveloped. It was a philosophy of how to live. Explaining the city as a confluence of natural resources, he noted that the origin of those resources was the indigenous landscape, or non-urban areas, that surrounded the metropolis. Taking his departure at Times Square in New York, MacKaye envisioned the traffic flows that had created the modern city. From the Midwestern cattle ranches and slaughter houses in St. Louis, to the cotton plantations of the South and the iron mines of the Great Lakes region, urban areas like New York, MacKaye pointed out, were intrinsically tied to the American landscape.

But whereas most movement historically had been from countryside to city, by the early twentieth century the flow was being reversed. Using highways as corridors for settlement, cities developed into uncontrolled urban sprawl, spilling over into the countryside and ultimately threatening the very resources upon which they depended. Lamenting the disappearance of indigenous landscapes, MacKaye substituted the metropolitan ideal with a new kind of organic regional city, structured as a cluster of towns designed to be in tune with its surroundings. Through the establishment of population levees in the form of, for example "a common public ground, serving the double purpose of a public forest and a public playground," urban population growth could be channelled into areas where it was desirable and did not threaten the natural environment (MacKaye, *The New Exploration* 179). At the same time, it would allow people to experience nature in close proximity. It was a far reaching and highly ambitious proposal, but in spite of being generally well received by critics, the public ignored *The New Exploration*, and the book sold less than 700 copies of its first edition.

Construction on the Appalachian Trail was finished in 1937, in some part due to the assistance of the Civilian Conservation Corps, and in 1938 the federal government declared a two-mile wide area the length of the entire trail to be protected (Cutler 58). Since it had first been suggested in 1921 the character of the project had changed from a pastoralized but essentially socialist ideal into a space that relied on the exclusion of human structures and influences. Instead of farm and food camps that integrated work and play in scenic surroundings and made up the basis for relocating part of the population, the actual trail only offered primitive shelters where hikers could seek refuge for the night. There was little to set the trail apart from the adjoining wilderness, and it was precisely wilderness that hikers came to experience.

There are of course several reasons why MacKaye's vision did not become reality. The prevailing popular and political sentiment at the time of the trail's birth was not in favor of the community-based reforms that MacKaye was proposing. One can only speculate as to what the trail might have looked like had MacKaye presented his idea a decade later to a reform-hungry New Deal audience. But the part of the project that did engage the popular imagination was the historically loaded preservationist discourse of a rapidly disappearing wilderness. MacKaye's original plan, as Marc Lucarelli notes, was "half pastoral elegy and half policy proposal," and the policy was ignored by all but the most fervent reformists (210). In spite of the trail's original utopian character, it was the rhetoric of recovering a romanticized wilderness past which became a rallying cry for nature enthusiasts all over the East Coast.

Three

A “machine-driven Arcadia”[2]: The TVA and the Transformation of the Tennessee Valley: 1933 to 1942

I

The creation of the Tennessee Valley Authority in 1933 provides a wonderful example of how the middle landscape was imitated in a variety of designed spaces for purposes ranging from the demonstration of planned communities to the creation of a regional tourist economy. Through the construction of a series of dams the TVA transformed the once unruly Tennessee River into a source of cheap hydroelectric power, creating jobs and stimulating the economy in the process. Headed by the visionary Arthur E. Morgan it was clear from the beginning that Roosevelt wanted a multi-purpose project with a more holistic approach to regional development. Consequently, the agency initiated a number of programs that, taken as a whole, promulgated a middle landscape ideology by seeking to reconcile nature and culture.

The scale and diversity of the TVA’s many projects allows the Authority to be studied in the context of a number of often overlapping historical narratives, and over the years the TVA has indeed become the focus of numerous memoirs and academic studies. Three directors—Arthur Morgan, David Lilienthal, and Gordon Clapp—have all published books about their years in the agency, and numerous academic studies have focused on the Authority in relation to topics as diverse as regional planning, navigation, engineering, architecture, organizational structure, economics, race, conservation, and politics. But the agency’s merger of economic and environmental interests to form a middle landscape for the future has not been given the attention it deserves. The TVA approached its role as environmental steward by seeking to create an environment in which the

---

[2] Hargrove, *Prisoners of Myth* 21.

utilization of natural resources was guided by a sustained-yield philosophy that became an integrated part of its corporate identity. Still today, the agency's role as environmental steward remains a highly visible part of its public profile. On the Authority's home page, for example, a heading reads "Energy, Environment, Economic Development," seemingly placing environmental matters ahead of economic ones. Another headline emphasizes an initiative called "Green Power Switch" which promotes alternative power sources including solar, wind, and methane gas. Of course, such projects must also be seen in light of the TVA's continued reliance, in the post-World War II era, on nuclear power, as well as its failure to upgrade outdated and polluting technologies at some of its coal power plants in order to minimize their harmful emissions (Devine 147-148). Yet even if often symbolic, the TVA's continued focus on achieving a balance between technology and nature has remained a central issue that points back to an ideology adopted in the first decade of operations.

The initial interest in the planned development of the Tennessee River had its roots in an old debate regarding ownership of the Wilson Dam located at Muscle Shoals in northern Alabama. A formidable sixty mile stretch of rapids, Muscle Shoals was by many considered the best site for the production of hydroelectric power in the eastern third of the country. But it was not until World War I and the demand for an increased domestic production of nitrates for ammunition that plans were finally made for the construction of a power plant there. In 1916 the federal government purchased the land around Muscle Shoals and two years later work on Wilson Dam was begun. The dam was finished in 1924, but because the war had ended before the power-generating facilities were complete, the project was abandoned and the plant was used only for the production of nitrate fertilizers for agricultural use. When the war ended, the Wilson administration tried to sell the land back to the private sector, and there were several bids to take over the plant. In the early 1920s, for example, business tycoon Henry Ford made repeated offers to buy the nitrate plant and lease the dam. Ford was against the centralization of power, be it public or private, and he proposed a plan in which the hydroelectric power generated at Muscle Shoals would help produce cheap fertilizer, electrify local farms, and stimulate the growth of small industry in the area. He even talked of creating a modern city, an announcement that brought land speculators flocking to northern Alabama (Talbert 72).

Ford's plan called for the creation of an agrarian utopia, but although he enjoyed widespread support in local farming communities where it was hoped that Ford might be able to turn Muscle Shoals into another Detroit, he could not get Congress to accept his proposal, and in 1924 he finally withdrew his offer.

It may seem odd that Ford could not win over Congress, considering that his proposal was reminiscent of the same desires that inspired the eventual creation of the TVA, but Ford's failure was not caused by political resistance to his plans. His inability to convince Congress was partly due to the work of southern utility companies who did not want the extra competition, and conservationists who feared that his plans would violate the Federal Water Power Act. The main resistance to his proposal, however, came from a number of Progressive congressmen who believed that private ownership of natural resources might ultimately lead to the nation's being held hostage by private enterprise. Instead, they proposed a federally controlled power program involving Muscle Shoals and the Tennessee Valley. Spearheading their bid for Muscle Shoals was a Republican Senator from Nebraska, George W. Norris, who had long been an advocate of public power development. Along with other prominent Progressives, including the former Biltmore forester and conservationist Gifford Pinchot, Norris secured widespread support for his ideas of cheap federally produced power and eventually gained Congressional approval for his project. But his proposal was vetoed twice by the Republican presidents Coolidge and Hoover, leaving the Muscle Shoals complex incomplete, "a symbol of the deadlock between corporations and advocates of a government power program" (Nye, *Electrifying America* 299).

If the public and political mood of the 1920s did not favor Norris' plan, the realities of the Great Depression made the American people more susceptible to federal intervention. But it was not until Roosevelt took office that Norris made any real headway with his Muscle Shoals project. Hoover's attempts to combat the Depression by cutting taxes and boosting private business failed, and in spite of his strategy to build public works to create jobs, his aversion to federal spending did not bode well for Norris' project. Norris strongly felt that Roosevelt would be the better choice for the common man, and he also correctly guessed that Roosevelt would be much more likely to back his proposal. In a controversial move, Norris broke with the

Republicans in order to back Roosevelt's candidacy. His decision meant that he had to leave the party altogether and run as an independent in the Congressional election of 1936; a sacrifice which testifies not only to the faith Norris must have had in the benefits of the Muscle Shoals proposal, but also to his belief in the Progressive approach to resource management.

Like Norris, Roosevelt was interested in government-sponsored regional planning and development. As Governor of New York, he had initiated "a state-wide planning movement to be based on a study of the proper use of the 30,000,000 acres of land in the state" (Roosevelt 123). He had also, according to his own notes, been interested in the development of the Tennessee Valley long before becoming president. Roosevelt's sympathy for the plight of the Tennessee Valley residents was the result of his many travels to the healing waters of Warm Springs, Georgia. These trips had revealed to him the Southern problems of poverty, slow technological progress, and isolation. To Roosevelt, a planned development of the region's resources might help bring the area up to the same level as the rest of the nation.

In January of 1933, before Roosevelt had been inaugurated, Norris travelled to Muscle Shoals together with him and a group of experts. Although the two men were in agreement on the need for development in the area, it was obvious that Norris' plans for the site were much less radical than Roosevelt's. His ideas for Muscle Shoals included only power production, flood control, and navigation. Dedicated to the notion of public planning and forced by the severity of the economic crisis at hand, Roosevelt envisioned a lift not only for the Tennessee Valley but for the nation as a whole. Inspired by urban planners such as Lewis Mumford and Benton MacKaye, Roosevelt saw the TVA as an opportunity to create a modern middle landscape for the rest of the country to follow. As Erwin Hargrove has noted, Roosevelt brought with him "long-standing interests in the conservation of natural resources, regional development, and the creation of urban settlements, all understood by him as means to a fresh start in developing an urban society that would harmonize the machine with nature" (20). Thus, when he wrote Congress on April 10 of that year Roosevelt spoke of a project which "if envisioned in its entirety, transcends mere power development; it enters the wide fields of flood control, soil erosion, afforestation, elimination from agricultural use of marginal lands, and distribution and diversification of industry .... a corporation clothed with the power of Government but possessed of the flexibility and initiative of a private enterprise. It should be charged with the broadest duty of planning for the proper use, conservation and development of the natural resources of the Tennessee River drainage basin and its adjoining territory for the general social and economic welfare of the Nation" (Roosevelt 122).

Fig. 7: Norris dam. Photo by author (2006).

Approximately one month later, on May 18, during the zenith of New Deal legislation, Congress passed the Tennessee Valley Authority Act. The act did not accommodate all of Roosevelt's far reaching visions, but specifically provided for the development of the river's hydroelectric potential, the creation of a navigation channel, a fertilizer production program, a reforestation program, and a flood control program. Sections 22 and 23 of the act granted the President the power to plan for the people of the Valley economically and socially, and to make surveys of the area in order to "aid further the proper use, conservation, and development of the natural resources of the Tennessee River drainage basin and of such adjoining territory as may be related to or materially affected by the development consequent to this Act, and to provide for the general welfare of the citizens of said areas." Roosevelt soon signed over that power to the TVA board, but the real significance of sections 22 and 23 lay in the continued possibility for a visionary approach to the development of the region in its entirety. The Tennessee River would become the engine in a social transformation that would encompass the entire region.

The ideas behind the TVA Act had been central to conservationist thinking as far back as Theodore Roosevelt's presidency. In 1908, Teddy Roosevelt's Inland Waterways Commission had concluded that any future development of the nation's waterways should take into account issues such as power development, flood control, and land reclamation. To ensure that the TVA's broad scope was not limited to theory, Franklin Roosevelt appointed Arthur Ernest Morgan as first chairman of the TVA board. A. E. Morgan had a reputation for being a socially engaged and visionary leader at Antioch College in Ohio. He was already engaged in a large water control project on the Ohio River, where he worked to combine river navigation and flood control with the creation of a number of recreational areas, a project that, in many ways, was similar to what Roosevelt had in mind for the Tennessee Valley.

Interestingly, Morgan had not voted for Roosevelt, but had instead cast his vote for the pragmatic progressivism of Hoover (Talbert 81). Yet Roosevelt and Morgan seemingly shared a vision of what the TVA could do for the people of the Tennessee Valley. Roosevelt's reply to Morgan's objection that the President hardly knew him, and certainly not well enough to hire him on the spot, was that he had been reading Morgan's book *Antioch Notes* and that he liked his vision (Morgan 6). Furthermore, Morgan was admittedly pleased when, during his first meeting with the President, Roosevelt talked "not about dams or electric power or fertilizer but about the quality of life of the people of the Tennessee Valley" (7). The two agreed that the project was to be about more than politics and the generation of hydroelectric power, and in his memoirs Morgan emphasized the project's philanthropic roots as the reason why he accepted the job. But as the developments of the following five years would show, Roosevelt also was a player in a larger political game, which meant that his support of Morgan had its limits.

Morgan's early appointment allowed him a chance to influence the TVA bill as it was still going through Congress. After contacting Senator Norris, Morgan made sure that the TVA would not rely on the Army Corps of Engineers for the construction of the dams. To Morgan the creation of a local and dedicated workforce was central to his vision of a Tennessee Valley utopia, and bringing in an Army Engineer workforce would get in the way of that. Also, in an act that perhaps more than anything shows Morgan's idealistic approach to the TVA, when he discovered a plan to pay the chairman of the board more than the two other directors, he had it changed so all three directors received the same pay (Talbert 88).

To serve with him on the board, Morgan first contacted Canadian biologist, Dr. Harcourt A. Morgan who had earned wide-spread recognition by describing the life cycle of the Boll Weevil, a small beetle that frequently wreaked havoc on the cotton fields of the South. But H. A. Morgan's biggest asset was probably that his years as president of the University of Tennessee had earned him the trust of the locals who had come to consider him one of their own. Something like that could easily prove useful when trying to sell a federal program to a region that traditionally was less than enthusiastic about federal intervention in what was perceived as local matters. The inclusion of H. A. Morgan, it was hoped, would help soften the agency's federal profile.

The search for a public-power expert proved to be more difficult. Morgan wanted someone talented, but who also had the strength needed to deal with the private power companies. He eventually chose David E. Lilienthal, a young and ambitious graduate from Harvard's Law School who was working for the Wisconsin Public Service Commission. Lilienthal had some reservations about his own qualifications, namely that he was not a Southerner, knew nothing about engineering, and was too young, but these objections apparently mattered little to Morgan (Lilienthal, *The Journals* 33). In most respects Lilienthal was a good choice. He possessed both the skill and the strength needed for the job at hand, and like A. E. Morgan he was a visionary. Other than that, however, the two men had little in common. Lilienthal was a realist and did not share Arthur Morgan's view of the TVA as a demonstration of how to build a new and better America. Instead, he felt that the agency should fulfil its job description from a purely technological standpoint and then leave it up to the people of the Valley to make the best of it. Arthur Morgan's paternalistic top-down approach to planning was not compatible with Lilienthal's idea of a grassroots democracy.

From the beginning Harcourt Morgan sided with Lilienthal. Although he did not agree with the utilitarian approach, Harcourt Morgan saw A. E. Morgan's utopian plans as essentially undemocratic. Before long the two disgruntled directors had combined to force Arthur Morgan to accept a restructuring of the board of directors. Instead of the previous arrangement with Arthur Morgan as the head of the board, a troika structure was formed in which each member essentially had equal power, but in different areas of planning. Arthur Morgan, still officially the chair of the board, took charge of social planning and engineering issues, Harcourt Morgan controlled projects relating to agriculture and forestry, and Lilienthal was placed at the head of the legal department and power issues. Although the new structure was a temporary solution to the daily conflicts, it also weakened the agency's structural coherence. The three directors communicated less and too often ran their own projects without first consulting the others (Hargrove 39).

To some extent blame for the problems between the three directors must be placed with Roosevelt. The agency was created without a clear organizational credo or any sense of structure. Beyond the sketchy descriptions of the agency's social potential and overarching goals, the three directors had little to go by when they first met. Also, since the TVA Act did not establish a prioritized order with regard to many of the agency's operations there were constant conflicts of interest between the different departments (Hargrove 19-41). The lack of clear goals made possible different interpretations of what the TVA's role was to be, and because Roosevelt never met with the whole board but only with the individual directors, he gave them reason to believe that he agreed with them all respectively. Lilienthal, for example, successfully inscribed the organization in a myth about grass roots participation that would permeate the TVA's sense of itself for decades to come, and never doubted that he had the President's support. Arthur Morgan on the other hand, who had become a great admirer of Roosevelt and the New Deal, believed that he was pursuing their common vision for the valley up until the moment in 1938 when Roosevelt finally decided to side with Lilienthal and remove Arthur Morgan from the board. From an organizational point of view the TVA was off to a rocky start.

Certainly to people in the valley the managerial turmoil between A. E. Morgan and Lilienthal, as well as the lack of clarity in policies and goals, must have seemed a somewhat theoretical issue. Already within the first year a number of projects were initiated, including the construction of Norris and Wheeler dams, both of which were completed by 1936. Among the first programs were also the construction of a system of power-lines from Norris to Muscle Shoals, the creation of a plan for the sale of the hydroelectricity, the development of a fertilizer program at Muscle Shoals, and a plan for how best to distribute the technical assistance needed to help farmers revitalize their operations.

The electrification of the valley was a clear priority and had been from the beginning. Prior to the TVA neither electricity nor electrically-run appliances were common in the valley. In fact, technological backwardness seemed a characteristic trait of rural America in the early 1930s. While almost ninety percent of urban America enjoyed the benefits of electricity by the time Roosevelt took office, only ten percent of rural homes were electrified. The fact that the hydroelectric potential of the Wilson Dam had never truly been considered, suggests that electricity was something which many saw as an almost exclusively urban phenomenon. But polls made as early as the 1920s suggested that farmers were less likely to leave electrified and mechanized farms, and so, by electrifying the Tennessee Valley Roosevelt hoped not only to improve the quality of life for its inhabitants, but also to counter the widespread migration from country to city.

Because farm life was physically hard and involved working long hours without any guarantees of what the harvest would bring, it was believed that any means by which daily tasks could be made less strenuous and more efficient might also help prevent farmers from leaving the farm. Plowing the fields could be accomplished both more easily and faster with a tractor than with a traditional horse drawn plow, and household chores such as washing the clothes, a time consuming, hard, and tedious form of manual labor, became much easier with a new electric washing machine. New refrigerators allowed people to store food for longer periods of time, which in turn saved them time and money because they no longer had to go to town for groceries as often, and new water heaters ensured that hot water was available at all times without the tiresome effort of having to heat it first. Radios meant that families could now sit at home listening to music or news, and feel a whole new sense of connectedness to the world around them.

The changes brought on by electricity were not just limited to the tangible transformations of everyday work routines. The new lifestyle was instrumental in creating a consumer-based economy that turned many local residents into debtors as they overextended themselves financially. Farmers often did not have the financial means to install the expensive wiring in their houses or to purchase the new electrical appliances needed to benefit from TVA-produced electricity. To help locals purchase the needed appliances, the TVA set up the Electric Home and Farm Authority (EHFA), a consumer credit affiliate that could offer farmers cheap loans to help them purchase wiring and other new household appliances (Kline 151). In what was originally one of Lilienthal's ideas, the EHFA also made deals with producers of electrical appliances to keep prices low, and, if such agreements failed, was able to lower prices by buying large quantities of products which were then resold through local cooperatives and power stations at more affordable prices.

Issues of flood control and navigation were also central to the restoration of the valley. Prior to damming the Tennessee, frequent floods and droughts had left local farmers and businessmen little chance of success. Every year floodwaters washed away much of the topsoil from the fields, damaged houses, destroyed industrial equipment, and left residents and businesses financially drained. The levees that had been built in order to prevent floods were by no means able to withstand the river, and often even worsened the flood risk because "they cut rivers off from their natural overflow areas, and forced more water through narrower channels, increasing flooding downstream" (Andrews 165). And when floods were not wreaking havoc, droughts often left the river unusable for irrigation or navigation.

Controlling floods to prevent soil erosion and conserving water for irrigation was not enough to restore the often eroded and depleted farm lands of the valley; a fertilizer program was needed. The research and production of fertilizer was Harcourt Morgan's main area of responsibility due to his background as an agricultural scientist. He had been researching phosphate depletion prior to his appointment with the TVA, and had found that the soil in the Tennessee Valley was greatly lacking in the natural mineral. Consequently, the traditional nitrate fertilizer would not work. Instead, he focused his attention on the new and more experimental phosphate fertilizer. Along with the presence of ground limestone, phosphate helped produce and fix nitrate in the ground, transforming useless land into fertile soil. At the Muscle Shoals plant TVA scientists were able to invent a new and efficient electric furnace which could utilize phosphate reserves more efficiently than before, making the fertilizer cheaper and of a higher quality than had previously been possible. By 1935 the TVA had a plant capable of producing 150,000 tons of phosphate fertilizer a year and demand increased as the benefits were documented. Eight years later, in 1943, the Authority spent more than eight million dollars on the production of phosphate fertilizers, the construction of fertilizer plants, as well as the purchase of phosphate ore reserves from which the mineral was extracted (Lilienthal 44-45 & 81).

Fig. 8: Results of TVA Fertilizer, 1942. Courtesy of the Franklin D. Roosevelt Presidential Library and Museum, Hyde Park, New York.

In order to convince the tradition-bound farming communities to use the new fertilizer, a plan was devised to test and prove the phosphate's value on local fields. Local communities would each select a farm which, sponsored by the TVA, would then function as a test farm where the wonders of the new phosphate fertilizers and the new machinery could be demonstrated to farmers. As Philip Selznick writes, "the TVA adopted the idea of a test-demonstration program, wherein the scientific function of appraising the practical value of concentrated fertilizers would be combined with the educational function of encouraging individual farmers and their neighbours to try the new materials and spread their use" (130). In this way neighbors could witness the benefits of the new technologies without any risk to their own farms or fields, while becoming potential customers in the process. At the same time, the TVA could test the fertilizer's effects on soil with a low yield.

The fertilizer programs did succeed in strengthening the agriculture of the Tennessee Valley. In 1943 The Southwest Virginia Agricultural Association prepared a publication in cooperation with the TVA and the Agricultural Extension Division of the Virginia Polytechnic Institute to document these changes. *Building a New Dominion* examined the revival of old pastures and the creation of new ones in the region through the use of TVA-produced phosphate fertilizers. The lack of phosphates and lime had been documented by an Agricultural Experiment Station set up by the Virginia Polytechnic Institute. These deficiencies could now be mended through the new fertilizer programs. Relying on Experiment Station data, individual farmers' stories, and a community example, the study found that after only a few years farmers in the area experienced "20 to 40 percent greater yields of pasture and other crops" and an increased number and quality of livestock (5).

In order for the phosphate fertilizers to work, however, much agriculture had to be changed into sod crops, particularly pasture and hay, which grows closer and, unlike row crops such as corn and tobacco, protected better against a loss of both water and soil. As individual test farms became successful, neighbors and communities became excited about the new technology, forming community organizations that could promote and monitor the programs. In the first year of operations, more than 200 farmers participated in the program, and already by 1936 several non-Tennessee Valley counties in Virginia signed up for the program. By July of 1942, 554 Tennessee Valley farms participated in the program along with 771 from 50 other counties (The Southwest Virginia 8-10).

The new fertilizer program was not only about improving crops. The new revitalized pasture lands also heightened the aesthetic qualities of the landscape, bringing it closer to the pastoral ideal. According to *Building a New Dominion*, people "in towns and cities talk about the improved appearance of the countryside—of farms made more attractive, of better homes and greener pastures with thicker sod" (5). To emphasize the idyll of the new landscape, the publication was illustrated with several scenic photographs of farmers in pastoral surroundings. The last page features the biblical quote "He maketh me lie down in green pastures: He leadeth me beside the still waters. He restoreth my soul…." superimposed on a photograph of two men standing in a green countryside next to a running brook, with grazing cattle and a farmhouse in the background. The message was clear: the new pastoral landscapes, courtesy of the TVA, restored the American soul.

Although much focus was given to the programs involving the production and distribution of electricity and fertilizer, flood control, and navigation, the TVA simultaneously initiated a number of smaller, but nevertheless important, programs that were infused with the values of the middle landscape. Some, like the massive reforestation effort, were directly related to flood control and the fight against soil erosion, while others had to do with recreation, tourism, or even, as in the case of the town of Norris, elusive ideas such as planning for the ideal future community.

II

The multi-purpose approach that permeated the TVA's operations affected the entire Tennessee Valley, but one particular dam remains central to the history of the agency. Norris Dam was the Authority's first, and the programs implemented in the surrounding area present a unique insight into the visions and ideas that were so crucial to the TVA's early years. Serving as a "laboratory in regional planning", at Norris reforestation, wildlife programs, recreational developments and urban planning were practiced (McDonald and Muldowny 4). A host of other conservation issues were also demonstrated as examples of the landscape that the TVA was trying to create. In time some demonstrations were deemed unfit while others were adopted in other parts of the valley, but Norris is in many ways representative of the ideology that guided the TVA in these early years.

Most of the new conservation practices introduced in the Tennessee Valley grew out of the belief that environmental sustainability correlated with a strong regional economy. Like the rest of the South, the Tennessee Valley had been settled predominantly by immigrants from northwestern Europe, but the agriculture from those regions was poorly suited to the steep hillsides and rocky ground. As one TVA report speculated, it is "unfortunate that this Valley of steep hills and vales, with its good soil, excellent climate, terrific thunder showers, and winter rains, did not get its early agriculture from the Mediterranean countries ... [where they] have retained the soil of their mountains and hillsides with food-bearing trees through the centuries" (Hershey). As already mentioned, the new agricultural practices included the introduction of fertilizers and new machinery on the fields, but in many cases traditional agriculture was also transformed into less destructive types of farming better suited to local conditions. The test farm system that had been used to demonstrate the new fertilizers was also used to promote these new farming practices, including crop rotation, small grains, fruit crops, forest tree crops, the raising of poultry, as well as dairy and sheep farming. Studies showed, for example, that a variety of nuts, including walnuts, hickory nuts, persimmon, papaw, and honey locust, were valuable and suitable crops for areas where agriculture and dairy could not be practiced; and allowing mast to grow afforded cheap feed for pigs and different species of wildlife. Also, such practices did not exhaust the soil as plowing the land did, and as traditional fields gave way to green pastures they dramatically changed the appearance of the landscape.

By diversifying farming methods in the valley the TVA improved the quality of the soil and with it the economic situation of many farmers. In order to help farmers from valleys that were too small for wheat and corn crops, a program was initiated to help them grow beans instead. Not only were beans highly nutritious, but they enriched fields with much needed nitrogen, thereby helping to maintain a healthy soil quality. The program was begun by Arthur Morgan after he received 300,000 federal relief dollars to spend as he saw fit. Always the visionary, Morgan felt that instead of handing the money out as immediate relief, a bean growing and canning operation could potentially help the poor mountain communities on a long term basis. Helping the farmers establish the Tennessee Valley Associated Cooperative for vegetable canning and hiring an experienced manager who could assist them, the venture proved a success until after the war when a major contract with the US military fell through and bankrupted the program (Morgan 72-73).

The restructuring of agricultural practices was based on the idea that agriculture had to be adapted to the surrounding environment in order to be successful. This meant taking the land into consideration when planning for a farm, a principle that broke with the age-old narrative of taming the wilderness and dominating the soil. In order for such efforts to work, a large-scale reforestation program was initiated in order to help prevent floods and soil erosion. Decades of widespread deforestation had left the soil barren and unprotected and unable to absorb the rainwater, which meant that the fertile topsoil was being washed off the fields and into the river whenever it rained, reducing the fertility of the fields and forcing farmers to once again clear new land in order to feed their families. It was a destructive cycle which, according to Lilienthal, could be broken only by recognizing that "in any valley of the world what happens on the *river* is largely determined by what happens on the *land*—by the kind of crops that farmers plant and harvest, by the type of machines they use, by the number of trees they cut down. The full benefits of stream and of soil cannot be realized by the people if the water and the land are not developed in harmony" (Lilienthal, *TVA* 50). It was estimated that by 1930 around 70% of the farmers in the Tennessee Valley were living on land with a rapidly diminishing resource base, and that 85% of the valley's 13 million acres of cultivated land was damaged by soil erosion; in fact, an estimate from 1934 by forestry and agricultural leaders had set "the reforestation need of the Tennessee Valley at 1.9 million acres" (Artman).

Some of the TVA's critics, however, felt that the claims of soil erosion had not been sufficiently proven and that the real worry of the agency's engineers was the accumulation of silt in their reservoirs. Other reservoirs had been known to fill up with silt from soil erosion, which minimized their storage capacity and thereby also the dam's ability to produce electricity. To avoid such problems, critics argued, the TVA promoted reforestation as well as a more pastoral use of the land instead of actual farming which would have left the earth exposed (Davidson, *The Tennessee – The New River* 290). Critics of the land restoration program were not wrong in identifying the TVA's interest in avoiding costly clean-ups and repairs by keeping the reservoirs free from silt. But clear-cutting had left farmlands in large parts of the Tennessee Valley in an increasingly poor state, and the planting of grass, trees, and bushes on the mountain slopes bordering the river and the reservoirs meant less silting.

In retrospect it can be hard to fathom the scope and complexity of the TVA's reforestation efforts. The program was placed under the jurisdiction of the Forestry Department, which was further divided into a number of specialized units including the Forest Land Administration, the Forest Tree Crop Unit, the Biological Readjustment Unit, and the Forest Education Unit. After being surveyed, the land was divided into two major categories: that which was best suited for private or corporate ownership and public lands. Most of the agency's programs were directed at the public lands which were divided into five subgroups: existing woodland or areas that were best suited for forest; land best suited for parks due to historic or scenic importance; recreational areas; protection areas, which denoted land where soil erosion and flood control issues were the most important; and finally land that was necessary for controlling public resources, including navigation, flood control, and power development. Although the different categories often overlapped, the reforestation efforts were primarily focused on forest lands and protection areas respectively. Park lands and recreational areas had to be more open and accessible, and the areas deemed necessary for controlling public resources had to be developed in accordance with other TVA departments such as the Division of Land Planning and Housing, the Division of Agriculture, and the Division of Engineering and Geology (Richards, "A Forest Policy").

In order to strengthen the new landscape, the TVA's forestry experts set out to test all known varieties of forest tree crops, as well as to locate new and hardier species within the valley. The tree nursery at Norris Dam became the center for the Forest Tree Crops Unit's development of new strains and crosses of trees and shrubs, as well as for the propagation of seedlings. The hardiest species were then planted on test grounds, after which the best varieties were brought into large nursery grounds where the different seedlings were grown before they were replanted in selected areas. To maximize the economic potential of the tree crops program, thousands of cross-bred nuts and seeds were also planted each year and set out in either forest plantings or in test blocks. Among the considerations regarding these crops were not just their market potential, but also their functionality as food for hogs, chickens, and wildlife. In fact, specific hog and chicken feeding programs were tied to the new tree crops experiments, just as balanced plantings were made to attract "insectivorous song birds … [to] save the woodlands, forests, and farm crops of the Tennessee Valley from insect attacks" (Hershey).

To plant the thousands of new trees, in 1933 the TVA was assigned workers from the Civilian Conservation Corps. Within three years a total of 28 CCC camps were divided among Virginia, North Carolina, Alabama, and Tennessee. The CCC workers were assigned to the Division of Forestry but were fed, clothed, and paid by the U.S. Army out of Emergency Conservation Work funds. Apart from planting the new seedlings, the CCC workers engaged in a number of activities in preparation for reforestation, including the building of check dams, the protection of hundreds of acres of eroding lands by covering them with a mulch of straw or a matting of brush—a technique that kept the rain water from immediately running off the barren land—and the digging of miles of diversion ditches that were to divert the run-off from areas where the danger of erosion was critical. All of these efforts were a necessary part of preparing the soil for the planting of the new seedlings (Tennessee Valley Authority, "TVA - CCC Camps").

Furthermore, at the Norris Forest tree crops nursery a series of interconnected programs were initiated. Different test plantings were created around the reservoir to prove how well different species responded to the conditions present in eroded lands, how fast they grew, how productive they were, and how well they supplied food and shelter for different kinds of animals. A number of special projects were also placed within the 117,000 acres of the Norris Lake Forest, including a permanent sample plot area of 350 acres, a demonstration planting area of 350 acres, a watershed runoff experiment area of 1700 acres, a demonstration fuel wood cutting area of 30 acres, and an area of 20 acres with virgin timber stands (Richards, "Project over the Valley" 3-4).

All of these demonstrations were designed to show how best to achieve sustainability in a future middle landscape, but programs were also initiated to educate inhabitants in the valley on how best to care for the new resources. One program in particular closely imitated the pastoral. Setting out to provide part-time employment for forest workers, it involved offering local workers a house on TVA lands complete with a small garden and some pasture. In return the workers were trained in forestry matters such as detecting, preventing and controlling fires, constructing trails, and planting new trees and shrubs. By paying the workers by the amount of work they did rather than employing them full-time—the workers were guaranteed 800 hours a year at a rate of 45 cents per hour—the program offered a seemingly inexpensive way of acquiring skilled forestry labor. At the same time, the arrangement allowed participants to maintain supplementary incomes, such as working on the malaria control program, through their farming and grazing rights, or even by offering recreation related services (Frank). By 1937 there were 22 trained workers living on TVA lands, constituting, according to chief forester Edward Richards, "the first forest community demonstration of its kind operating in the United States" ("Projects over the Valley" 3). Beneath the obvious economic benefits of the program, the underlying vision of the forest community demonstration was one of showing the public how to create a society in balance with its surrounding environment.

In a sense, the reforestation program expressed a recovery narrative in which denuded, eroded, and otherwise abused lands were reclaimed and made a part of the TVA's pastoralized landscape, and such efforts were not limited to public lands. From a TVA point of view, the future of the program depended on the willingness and ability of the population to engage themselves in the conservation project. Anticipating Aldo Leopold's land ethic and his proposition to extend "the social conscience from people to land," one TVA report noted that the "Work of Forestry Education must not concern mere physical or material values. Forestry fails if it does not bring beauty, inspiration, and an appreciation of the powers of nature with the obligation of man to work in harmony therewith" (Leopold 209; Perry 18). The Forest Education Unit was charged with the general education of the people in the Tennessee Valley in forestry-related matters, including information on how to prevent forest fires, material for the media and radio speeches, informational posters and literature, and even essay contests for grade and high school students.

One way to engage the local population in the reforestation program was by convincing them of how much they stood to gain from participating. Even in its depleted state the region could potentially produce an estimated 52 million dollars worth of raw materials annually, which was enough to wholly or partially sustain 250,000 farm families and 100,000 non-farm families (Thurmond). The restoration and conservation of the forests could potentially double or triple local employment, income, and business.

Of course, the development of private lands happened in cooperation with the owners. Private and corporate landowners supplied the manpower and tools, while the TVA supplied the trees and the technical assistance for planting them. The development of private lands increased after 1942 when the CCC program was ended, but it still centered on the recovery theme. Because of the war the public work force was gone, and the continued reforestation of the region suddenly depended on private initiative. In a 1941 manuscript for a promotional and educational film titled “Waste Lands to Woodlands” the TVA’s Forestry Relations Department shows how a program was established to help such private efforts at forestry. Interestingly, the narrative romanticized not only the changes that had already been brought to the region, but also the balanced landscape ideal that permeated the agency’s rhetoric.

Divided into four sections, the “Waste Lands to Woodlands” manuscript opens with “A Valley Scene.” Showing idyllic images of livestock grazing on an “improved pasture,” a “farm house with woodland in rear,” or a “farmer stacking firewood” the narrator educates the viewer about the importance of such private woodlands, and how best to keep them healthy and profitable. As the narration continues, the images initially show sawmills in operation and lumber yards, and then, as the first part of the movie ends, footage of eroded and over-grazed pastures and severely eroded hillsides.

The second and third parts of the film focus on the scientific recovery of the pastoral landscapes that were lost. It first shows TVA-sponsored tree nurseries where thousands of seedlings are produced, and then a number of success stories such as Tishomingo County, Mississippi, where 150 landowners planted 324,000 seedlings on 275 acres of eroded farmland, and Dekalb County, Alabama, where more than 100 landowners planted 400,000 trees on 315 acres of exhausted land.

The fourth and final part of the film begins with an image of a ruined landscape which is soon replaced by that of “pine seedlings waving with the wind”. The narrator praises the combined efforts of the federal agency and the local farmers to begin successful private timber production while preventing future soil erosion. As the narrator informs potential woodland farmers about how best to manage their crop, the final image is a romantic one of pine trees against the sky.

Relying heavily on images of either wasteland or pastoral scenery, the “Waste Lands to Woodlands” manuscript adequately captures the central theme of the TVA’s reforestation effort. By planting trees the agency reclaimed the land for the benefit of both man and nature. Framed by images of a landscape in the process of being recreated, the Tennessee Valley presented in the manuscript is not one dominated by industry or the urban markets where much of the wood was subsequently sold. Rather, "From Waste Lands to Woodlands" presents a vision dominated by scenes featuring the woodland farmer of the future, cultivating his crop in balance with his surrounding environment, embodying the kind of pastoral message that has been so central to the American imagination.

Along with the reforestation program, the project that most embodied the middle landscape vision was the construction of the small town of Norris. Dubbed a "new sort of town, with architecture blending the utilitarian with the artistic in pleasing harmony," Norris was an attempt to create an entire community in dialogue with its surrounding environment ("New Government Sponsored"). Designs and building materials, even aspects of social planning, all came together in the creation of a town meant to blend in with the landscape. Although described by McDonald and Muldowny as a "suburb in the wilderness," Norris was in fact a planned community located in the midst of a sculpted and highly pastoralized environment (217).

Located approximately four miles from the dam, Norris was originally built to house the more than 2000 construction workers employed there, but planners soon realized that the creation of temporary accommodations for the many workers would require almost the same kind of infrastructure as a permanent community. Also, since most of the workers were local and would stay to work at the new power station or in other industrial jobs after the completion of the dam, the TVA decided that the money was better spent on creating a permanent village.

Part of the motivation behind the creation of Norris was the TVA's attempt to combat the real estate speculation that threatened the agency's democratic vision by raising prices. By building the houses, the Authority was able to eliminate contractors and cut costs to approximately one third of the price, which allowed workers to rent a house for as little as 14 to 25 dollars a month. One reason the TVA was able to keep construction costs low was that in December of 1933 500 Civil Works Administration (CWA) workers were allocated to the Norris project to construct roads and dig ditches for sewers ("500 Get Jobs at Townsite"). Before the program was cancelled only one year later, in early 1934, the CWA employed up to four million people nationwide in different types of jobs for an average pay of 15 dollars a week. Just like the CCC camps that were employed to beautify roads, create trails, and plant trees, even if only for a brief period, the CWA workers became a welcome source of inexpensive labor.

Private land speculators were understandably angered by the competition from the federal government. How were they to compete with a federal agency that was able to draw on national relief aid funds in order to keep prices low? As had been the case with the debate over the Authority's role on the power market, accusations for being anti-capitalist and for undermining the real estate market were soon directed at the TVA. And critics were right in pointing out that building federal housing at low costs was anti-competitive, but as defenders of Norris pointed out, the criticism also had to do with the unease that was created in the face of such social reform. The liberal intellectual journal *The Nation* commented: "Had the Authority decided to erect on mud flats long rows of ugly clapboard or yellow-brick houses … the only criticism would have come from unimportant liberal circles. But nothing in the world will save the Authority from a volley of dead cats for producing for common workers a charming town, which has even been located with the desire to provide for every dwelling soul satisfying vistas of the beautiful misty hills" (Amend 645).

Planning for Norris began before the very first TVA board meeting had taken place. During their initial meetings Roosevelt and Arthur Morgan had discussed the creation of a model town instead of the usual temporary quarters used to house workers. Responsibility for the creation of the town rested with the Division of Land Planning and Housing which was headed by Earle Sumner Draper, a visionary who shared many of Arthur Morgan's ideas. Draper was a landscape architect by education, and like Olmsted, whom he greatly admired, he believed that developments should be respectful of the land and be in tune with their surroundings. Consequently, as Draper noted in a 1933 article for the *Chattanooga Times*, the houses in Norris were to be designed as "a natural outgrowth of the very soil, the forests, the mountain slopes of which they are a part" (Draper, "Authority Plans").

The town's design emphasized using local materials and regional architecture. Several different types of houses were available in Norris. The most common model had three or four rooms and was designed to house a normal family, but homes ranged from a bachelor-sized two room domicile to houses with seven rooms for larger families. The exterior walls were built with local brick, variegating in color and texture, and each home had its own vegetable garden. But such an idyll was not a goal in itself; rather the new town was desirable because it was practical. As Draper wrote, if "there's idealism in these house designs, it is of a practical sort rather than a romantic. There are no pseudo-Old World houses with lath half-timberwork ... New England quaintness or Mediterranean cleverness" (Draper, "Authority Plans"). More than one hundred cinderblock houses and a few apartment buildings were also built for those workers who could still not afford to live in the stylized and regionalized housing project, but the majority of houses in Norris were constructed "specifically to conform to a mountain setting, rugged and unspoiled; to suit mountain climate, capricious but mild, and above all, to serve a mountaineer culture during and after a period of unmeasurable change" ("Homes of Workers").

Fig. 9 : Housing for TVA workers in Norris, TN, 1933-1945. Courtesy of the FSA/OWI Collection at the Library of Congress Prints & Photographs Division, Washington, D.C. 20540 USA.

The undeniable potential of a project like Norris naturally engaged and excited some of the agency's most distinguished planners. Among those involved were Earle S. Draper, Tracy Augur, and Carroll Towne who were in charge of the complete development of the town. Even Benton MacKaye, who was employed as a regional planner by Draper in 1933, but who never managed to fit in with the TVA, became involved in the project. Many of the ideas that came to define the town originated in the garden city movement that had proven such a major inspiration for the members of the Regional Planning Association of America, but Norris went beyond the garden city ideal (Schaffer, "Norris" 3). Even Clarence Stein and Henry Wright's recent Radburn experiment in New Jersey seemed limited in comparison. The TVA town was about social planning and the kind of truly visionary urban reform project that theorists like Lewis Mumford had written about years earlier.

Recognizing that land use and settlement patterns were central to social issues, planners approached the problem from a holistic angle, hoping that Norris would become the shining beacon and model for other towns in the region. Norris "emerged as an important demonstration not only because it incorporated industrial and agricultural activities into its economic base, but also because it visually represented the land use patterns which planners hoped to replicate throughout the valley—tightly knit, relatively self-contained communities surrounded by undeveloped green space as a means of achieving both physical definition and environmental balance" (Schaffer, "Environment and TVA" 353). No property lines existed, which meant that fences and hedges were useless. The result was an open and almost park-like democratic milieu (Cann 33). Open spaces where nothing could be built were an integral part of the town plan, and the town was furthermore surrounded by a greenbelt of both undeveloped nature and agricultural areas that were designed to protect it from the kind of unplanned urban and suburban expansion that had marred so many other American urban areas. Recreational spaces were created within the city as well as in the greenbelt, and many residential streets were dead-end to keep through-traffic out.

In order to promote small-scale agriculture, four to five acre plots of land were located nearby for anyone who should desire to do a little farming as a supplement to their everyday industrial work, and an agricultural center was built in the middle of town for local farmers to sell their goods. Arthur Morgan introduced a system by which workers only had a 33 hour workweek, and were offered classes in different crafts during their free time. The system should ideally help combat unemployment, while stimulating the production of different handicrafts such as wood-works and ceramics in an effort to make home-industries a part of the community's economic base. By decentralizing trade, planners believed that more autonomy was given the individual communities, which in turn would help them thrive. Just as adults in Norris were offered training in crafts, children were educated in skills that might prepare them for life in the new middle landscape. In the local school—which functioned both as a library, a town hall, a church, and a movie theatre (Gray 33) —students from seventh through twelfth grade were given training in experimental classes such as "agriculture, home problems, and in the mechanical, woodworking, metal working, electrical and automotive shops" (Cann 40).

By 1937, however, the TVA board decided that Norris no longer fit the agency's program. Arthur Morgan was on the verge of leaving and Congress was openly criticizing the Authority for the project which they felt was far from economically viable. Consequently, preparations were made to sell the town. Many houses in Norris had remained too expensive for ordinary workers, and as a result the community attracted mostly higher paid personnel. Also, few non-TVA personnel had moved there and the local business community did not thrive. The town had failed to develop independently of the TVA, and although it may be argued that Norris was never really given the chance the project deserved, Congress focused on the fact that the town had proven unable to sustain the original vision.

The war in Europe and the resulting shift from multi-purpose planning to power production, postponed the sale of Norris until 1948 when it was sold to a consortium of Philadelphia businessmen for 2.1 million dollars, and subsequently resold to the individual tenants. Balancing nature and culture, industry and agriculture, Norris was, at least in theory, a perfect example of the pastoralized middle landscape. It attempted to create a modern version of the original country hamlet, mimicking a pre-industrial social structure in which production was decentralized and grounded within the community. By simulating the pastoral ideal the TVA's urban planners sought to do away with both social injustices and modernity, creating a utopian community inspired by pre-industrial America. As we will see later in this chapter, however, the utopianism of Norris was both highly politicized and racialized, making an unintended statement about exactly what kind of future was being envisioned.

III

During the 1920s and 1930s new and improved means of transportation, along with the construction of better roads, meant that the American landscape was quickly becoming a playground to be developed for the people; and with the increase in tourism a utilitarian approach to nature developed that corresponded well with the fundamental characteristics of the middle landscape. That form of utilitarianism was reflected in the developments of the Tennessee Valley. It seemed obvious from the TVA's conception that tourism was to become a cornerstone in the new regional economy. The existence of no fewer than 85 golf courses in the southeastern part of the Tennessee Valley by 1938, many using the mountains as a dramatic backdrop for players to enjoy, confirmed that there was a market for such nature-based activities (Tennessee Valley Authority, *Recreational Development* xii).

Arthur Morgan and other TVA planners recognized the social and economic potential of recreation and tourism early on, and the development of the region as a tourist haven became one of the central concerns of the Authority's conservation efforts. Along with reforestation came parks and trails, and along with the new reservoirs came boating facilities and even small beaches. Even when deciding on land purchase policies, TVA officials had to take into consideration what areas had the largest recreational potential. The agency specifically targeted the growing number of tourists who, increasingly anxious about modernity, sought an escape into the non-urban. Tourist sites were created in the image of idyllic Arcadian landscapes that, it was believed, visitors wanted to experience. After all, as the Authority's Chief Forester Willis M. Baker noted in an internal report, people "seeking an outdoor playground are not attracted to a devastated region" (Baker, "Conservation").

By marketing the new reservoirs as wonderlands for angling, boating, and water-skiing, creating parks around the reservoirs to allow visiting families to picnic in scenic surroundings, constructing hundreds of miles of trails in the vicinity of the dams to accommodate hikers, and by carefully managing fish and game populations to make the new landscape attractive to a variety of nature enthusiasts, including ornithologists and hunters, the TVA responded to the growing demand for non-urban recreation. Along with the power programs and the navigation channel, the economic development of the region fostered a new kind of landscape that emphasized a union with nature through play rather than work. Whereas Benton MacKaye's vision for an Appalachian Trail was based on the idea that working with nature could be made into play, non-urban recreation was based on the idea of escaping work altogether.

Thus, recreation and tourism were not only about economic gains. Activities such as hiking, bird watching, boating, fishing, and hunting took on a social and almost moral dimension that in part was based on a critique of modern society:

> Our present form of civilization too often prevents us from finding varied recreation in the home and city, the two agencies to which one would logically turn. The *home*, in many cases, has shrunk to one or two rooms, with no private grounds around it, and the *city*, made up of thousands of these dwellings closely compacted, cannot provide a complete environment for the various types which make up the composite term *recreation*. There are activities such as reading, indoor games and various hobbies which the home can provide. Other activities can be provided for by the parks, playgrounds, theatres and amusement halls of the city. It is left to the non-urban areas, however, to provide facilities for many of the more valuable forms of recreation. Only in such areas can be found necessary space for such activities as hunting, fishing, hiking, and picnicking. Only in such areas can one come in contact with the rural and primitive environments so necessary to complete the background of human activity and to provide a genuine

> reanimation of the spirit. (Tennessee Valley Authority, *Non-Urban Outdoor* 2-3).

Visits to the great outdoors were invested with a sense of spiritual and moral purpose which the city and the modern home were seemingly incapable of providing. In the TVA's rhetoric, nature, even if planned and gardened, became an antidote to civilization, and tourism expressed the pastoral narrative in a new and commercialized form.

In order to accommodate such a reanimation of the spirit, the TVA established no fewer than three large parks around Norris Dam: Cove Lake State Park of about 800 acres, Norris Park of 3,887 acres, and Big Ridge Park which, with its 4500 acres, was the largest of the TVA parks. Norris Park combined the more untouched "natural study areas" with heavy-use picnic grounds and camping sites. Constructed between 1934 and 1937, the park, like the Norris Freeway, was the result of the agency's cooperation with the Civilian Conservation Corps, and it attracted many people from the eastern part of the state.

Fig. 10: Big Ridge Lake, 1933-1945. Courtesy of the FSA/OWI Collection at the Library of Congress Prints & Photographs Division, Washington, D.C. 20540 USA.

The reservoirs became central to the new recreational landscape. On their waters people boated, swam, and fished, and on the adjoining lands parks, camping and picnic sites, and trails were established. One lady with whom I spoke still recalled how, as a child growing up in Knoxville, she and her family would drive up to the new Norris reservoir for a day of swimming and picnicking. Such unified development was possible mainly because of Arthur Morgan's ability to push through a new approach to land acquisition. Traditionally, only the land directly affected would have been purchased, but during his five years as director of the TVA, Morgan consistently secured the title to all marginal lands for one quarter of a mile around each reservoir in order "to control the development of the area, especially of its recreation facilities" (Morgan 69). As one report expressed it, the unguided development of recreational resources tended to "produce short-sighted localized exploitation" (Tennessee Valley Authority, *Recreational Development* viii).

The emphasis on controlling the recreational resources of the region was clearly stated in the "Revised Forest Development Section Policy Statement" from 1938. The report described any attempt at gaining a private monopoly over "public utilities and conveniences, or on public property" as "indefensible and intolerable," and further proposed that most requests to establish private campsites, cabins, and concessions should be denied. In fact, any permits given "to private parties for these purposes should be issued rarely and cautiously, taking every measure to insure that no legal vested right in the land or water can be claimed by the permittee" (Perry 6). Furthermore, the report went on to note, on the rare occasions when private developments were allowed, they should be only temporary arrangements that were renewable on a monthly basis, and they should be located in permanent structures owned and operated by the TVA. Had an entrepreneur been lucky and forthcoming enough to have gotten a lease to run a concession stand on TVA lands, there were still strict requirements regarding the standards he should meet as well as the kinds of service that he was allowed to provide.

Wildlife was also considered important in attracting tourists. Fish and game functioned as a source of income and food to many residents, but the protection and management of the local animal population was also central to the success of bringing tourism to the valley. The Norris area, for example, lay within driving distance of most of the eastern seaboard, which meant that the new recreational areas could potentially attract hundreds of thousands of hunters and anglers from places like Atlanta, Washington, Philadelphia, and possibly even New York. Thus, game and fish were carefully bred and released in order to promote such tourism.

To ensure that the new lakes were not overfished, the TVA actively monitored and studied the populations in order to protect and develop the new resource, learning, among other things, that lowering the water level in the reservoirs during spawning season could prove disastrous. In recognition of the potentially negative effects the new dams could have upon the local environment, the Biological Readjustment Unit was established in order to secure the wildlife populations. The unit was charged with securing the high standard of fishing, while drawing up plans on how to improve conditions for fish and game in the new landscape (Perry 11). In order to boost and maintain fish populations, for example, hatcheries were built at Norris Dam and on the Elk River in northern Alabama. Placed under the jurisdiction of the Biological Readjustment Unit, the hatcheries helped restock the fish populations. Meanwhile, an island on the Norris Lake Forest was set aside as a wildlife sanctuary, while two other areas were designated as natural study areas, and two more were reserved for game management demonstrations.

Fears about the reaction of different animal species to the new man-made environments were soon laid to rest as environmentalists and engineers alike realized that most species had little trouble getting used to the new landscape. In fact, fish thrived in the new lakes and soon began breeding, and the reservoirs also began attracting different kinds of migratory birds. Even the mussels that had been such an important part of the old river and the regional trade, and which many had believed could not survive the shift from a river to a lake-type environment, managed to survive the changes. As the wildlife populations of the Norris Dam area began to thrive, the TVA began marketing the recreational possibilities of the region. The program's success was apparent: in June 1940, a mere seven years after the TVA had begun its operations, the Norris Reservoir supported around 57,000 fishing trips (Eschemeyer). It is uncertain whether the name "Great Lakes of the South" was actually thought up as a part of the promotional campaigns, but whatever the case the agency was quick to utilize such comparisons with another of America's favorite recreational landscapes (Eschemeyer). Furthermore, the name also lent a certain natural feel to the man-made lakes.

Yet while the protection of wildlife meant that some animal species increased through breeding programs, other species such as wolf, elk, and buffalo, that had become extinct because of indiscriminate hunting practices and the alteration of their natural habitats, were never reintroduced (Tennessee Valley Authority, *Recreational Development* 8). One can only speculate as to the reasons why some species populations were boosted while other native species were allowed to remain extinct. Perhaps reintroduction programs were simply more labor intensive and costly than ordinary breeding programs, or perhaps the potentially dangerous animals were not deemed compatible with the utilitarianism of the new middle landscape. This less than natural selection emphasizes how the new landscape was in fact a social construction whose features were carefully manipulated in order to appeal to the widest possible range of visitors.

Planning for modern tourism also meant recognizing the role and importance of the automobile culture, as well as other modes of transportation. As a TVA report on possible recreational developments expressed it, whether "for better or worse, the automobile and improved roads have made possible a mobile type of vacation which has proved immensely popular" (Tennessee Valley Authority, *Recreational Development* 15). From April 1934 to November 1935 two CCC camps were set to work on improving and beautifying the Norris Freeway. Their work included "seeding and sodding, bank sloping, rough and fine grading, ditching, planting, wayside picnic areas" (37). A 250 foot wide right-of-way was established for the construction of the road, where no filling stations, billboards, or other types of constructions were allowed. In every detail, the road was designed to give the impression of being "molded into the earth" (Draper, "The TVA Freeway" 20). In his two-volume history of the Tennessee River, Donald Davidson describes the approach to the new Norris Dam:

> One came to Norris Dam from Knoxville and after following a state highway turned on to the Norris Freeway. The Freeway led the motorist along a pleasant valley. On either side, as the road wound among the ridges, the motorist could see, or begin to see, neatly

> fostered illustrations of the forest conservation and proper land use which the authority was committed to sponsor. The nearer he came to Norris Dam, the more the countryside took on the appearance of an amiable wild park which told him, without words, how Tennessee ought to look if it were benevolently protected from man's foolishness. (Davidson, *The Tennessee – The New River* 229)

Allowing the road to follow the contours of the landscape, the TVA's planners clearly saw the Norris Freeway as an organic structure that blended with the surroundings rather than a utilitarianism of transportation.

Winding roads overlooking spectacular and varied scenery were important to the new mobile tourists, and as such also became central to the TVA's planning efforts. But the immense popularity of automobile-based tourism also meant that the tourists had to be protected from themselves. While it was relatively easy to ensure beautiful lookout points along the roads and to allow motorists to marvel at the landscape, it all demanded keeping the roads free of billboards and other such structures. As one report notes: "Without suitable regulation, they [the roads] cannot be expected to remain so as tourism increases" (Tennessee Valley Authority, *Recreational Development* 15).

Other modes of transportation were also included in the tourism strategy. Plans for hiking, inspired by the Appalachian Trail, and boating, the result of the new reservoirs, became part of the new landscape. The river was already a favorite among vacationing houseboat owners, and the number of private river cruises was only likely to increase on the new river, especially when access roads and docks were built. Even the possibility of pleasure-flying, which was expected to become a recreational activity of the future, was taken into consideration as the new landscape was being thought out. The dam itself also became part of the dramatic scenic spectacle. Man-made viewpoints along the reservoir allowed for wonderful views of the impressive concrete structure framed by the blue of the reservoir and the lush green landscape, much of which the TVA had planted. The dam in itself became part of the spectacle, demonstrating how, in the future, technology and nature would complement each other.

All the while, the TVA had to be careful how it presented itself to visitors. Through the new tourist landscapes, the Authority could inscribe itself into a narrative that was appealing to the modern consumer but which still incorporated the pastoral trope. One way to achieve this goal was by including displays of the distinct regional culture including folklore, ballads, handcraft, and architecture. Often such displays tended to emphasize the rusticity of the new landscape, adding a sense of history and authenticity to an otherwise sculpted and gardened space. At Norris Dam, for example, a pre-industrial gristmill was displayed in the vicinity of the new hydroelectric dam. The closeness of the two structures symbolically integrated past and future technologies in a utopian vision of coexistence.

Fig. 11: The Rice grist mill in Norris Dam State Park. Photo by author (2006).

Originally constructed in 1798 by a settler named James Rice, the gristmill had been located on land that was to be flooded by the Norris Dam. When the TVA purchased the Rice land in 1935, the mill was disassembled with the help of the Civilian Conservation Corps and the National Park Service. Each part was carefully labelled so that it might be rebuilt and displayed. Through the years the mill had played the role of power source for both a sawmill, a cotton gin, a trip hammer, and even a dynamo supplying electricity for the Rice household in the late nineteenth and early twentieth century. That role, however, was taken over by the massive turbines of the new dam standing less than a mile away. While the relocated mill was meant to signal a sense of balance, idyll, and respect for the region's past, its placement in the shadow of the dam also came to signify the obsolescence of an old technology that had been reduced to a working demonstration for the amusement of visiting tourists.

The TVA also used its tourist landscapes in different manners. By displaying an industrialized and ravaged area as a contrast to the Authority's conservation efforts, the landscape became part of a political argument. Thus, in the Ducktown district in the southeastern corner of Tennessee an industrial wasteland was used as an educational opportunity, showing tourists what had happened where the scales had tipped in favor of exploitation rather than conservation. One report described the Ducktown area in the following way: "The sulphurous fumes of the copper smelters in the Ducktown district of southeastern Tennessee have created a tourist attraction unique in its nature. These fumes have denuded the vegetation of the surrounding region, and the resulting loss of cover has laid the land bare to the ravages of erosion. This vivid exhibit of the need for conservation policy is made complete when the tourist realizes that the erosion results not only in the wasting of timber and land, but also in the silting of reservoirs and navigation channels" (Tennessee Valley Authority, *Recreational Development* 24-25). The barren and desolate landscape at Ducktown functioned to place the TVA's conservation-minded efforts into an environmental and economic context. It exemplified "unbalanced resource development," a sight so hideous that no one could "look upon this horror … without a shudder" (Lilienthal, *TVA* 53). Not only did examples such as Ducktown help shock the public and promote conservation, they undoubtedly also functioned as valuable tools in the constant fight against the agency's political enemies. After all, from a TVA perspective Ducktown was not an isolated case of environmental destruction, but, as Lilienthal put it, a "reflection of our national thinking" (54).

IV

Although a majority of the people in the Tennessee Valley either welcomed the TVA programs, or merely accepted the Authority as an unavoidable part of progress, there were also those who either resented the federal intervention and the shift in values that followed the transformation from countryside to tourist heaven, or who became marginalized as a result of the new landscape. Politically, the TVA's structure as a public corporation, its encroachment on an otherwise privatized power industry, and its highly philanthropic scope left the new agency vulnerable to criticism from many sides. Republicans called it socialist, un-American, and a clear-cut case of pork-barrelling; the power industry criticized it as being anti-competitive; local politicians around the country wanted an equal measure of federal support for their region; and Southern intellectuals like Donald Davidson regretted what they saw as "another Yankee raid into southern territory" (Davidson, "Where Regionalism" 27). Counter narratives are a part of any history, and the story of the pastoralization of the Tennessee Valley would not be complete without them to contradict and contextualize the ideology that the TVA was selling of a balanced and democratic environment. Among the most dominant are the stories of the many displaced people who were forced to leave to make way for the new landscape, as well as the problems experienced by African Americans who were not only discriminated against when seeking employment with the TVA, but who were also denied access to the new landscape.

Nowhere were the changes imposed upon the landscape more visible and dramatic than in the areas where the new dams caused the river to back up for miles. The water trapped in the large reservoirs flooded farms and entire towns as it rose. For the construction of the Norris dam and reservoir alone, 153,000 acres were purchased, displacing 3000 families (McDonald and Muldowny 4). A few years later, when the 206 feet high and more than one and a half mile long Kentucky Dam was constructed it flooded the two river towns of

Birmingham and Newburg, causing the relocation of 3,500 families, or close to 15,000 people.

From the very beginning the Authority was divided on the question of how to resettle the dispossessed. Considering the fundamental disagreements between the directors, it is hardly surprising that coming up with a relocation policy proved problematic. Proposals ranged from the creation of small model communities made up of ten to twelve resettled families, to letting farmers themselves figure out how to spend the money they received for their land. The issue proved hard to solve because it touched upon the basic question of what role the TVA was to play in the valley. Lilienthal wanted as little as possible to do with social planning and felt that people should be allowed to work things out for themselves once they received their settlement. Arthur Morgan, on the other hand, wanted to use the opportunity to further plan for the region. As time passed and no agreement was reached, a temporary office was created to collect and study the specific wishes of the families that were being relocated in order to cross-reference them with available farms in the valley. It was a temporary compromise, and far less ambitious than some of the other suggestions put forth, but it turned out to be a workable plan that allowed the individual families a sense of autonomy, and the arrangement soon became permanent.

Relocation was a formidable task, and it was not made any easier by Arthur Morgan's land purchase policy. Whereas Harcourt Morgan wanted to leave as much farmable land as possible, A.E. Morgan felt it important to make sure that no farmer was left with so little land that he would not be able to support himself and his family; a policy of minimal purchase, after all, might cut farms in half. The policy of overpurchasing ensured that the Authority was able to control the land adjoining the new reservoirs. Overpurchasing, however, also meant that a far greater number of people were being displaced than what would otherwise have been necessary. According to one estimate, for the building of Norris Dam alone the policy of overpurchasing may have caused as many as 1500 families to have been relocated without cause (McDonald and Muldowny 136).

In general, people were willing to move, and to many young farmers it was even an exciting step away from a social structure from which they had been unable to break free. A 1949 evaluation of the land acquisition program showed that only 4.4 percent of the relocated inhabitants at Norris had refused to sell. On average only 3.0 percent of the Tennessee Valley land acquisitions had proven problematic (McDonald and Muldowny 138). Naturally, such numbers do not reveal how many families were against moving but felt unequipped to fight a federal agency, but they do suggest that the TVA in general did much to ease the process. To take care of the people being relocated, for example, in each community an administrator was hired to ensure that families were offered new farms suited to their specific needs. Administrators travelled the countryside armed with federal money, interviewing and negotiating with individual families in order to persuade them to leave their homes and, when all else failed, calling upon the right of eminent domain. By 1946 the Tennessee Valley Authority had relocated no less than 13,449 families, an estimated 50,000 people, all of whom had specific wishes, histories, and traditions that were to be respected.

The types of problems which the TVA's administrators might be confronted with when trying to relocate families were many and complex. Whereas many young farmers saw the relocation program as a chance to break with traditions and try out a new and different way of life, older residents were much more attached to the land on which they had spent their entire lives. Many of the farms that were being flooded had been in families for generations, and such places were invested with memories of a long family history and years of hard work. Indeed, each family presented a unique situation, as the story of Erasmus Lindamood proved. Accepting that he had to leave his home, Lindamood insisted that the fire from his fireplace be brought with him to his new home. As he told a visiting reporter from the *Knoxville News-Sentinel* in the fall of 1934:

> I was born in the light of this fire seventy years ago. My father built this house in 1859. I won't swear the fire has never once gone out. I do know the ashes in that fireplace seldom, if ever, have got cold in my lifetime.
>
> Winter and summer that fire has been fanned up at night. It was covered day and night around. A few times I've been off from home two or three or four days at a time. But the old fire, she's been fanned up every night I was here—and I reckon when I wasn't.

> There's plenty of matches nowadays. I may be childish. But since I'm going to move out of here, see water lapping around the house and the crib and the smokehouse, and the chicken-house, it'll mean something to take the fire from the old house with me. (Watkins 162-163)

To people like Lindamood the home was all they had ever known, and some farms even contained family burial plots which had to be relocated along with the rest of the family.

Some of the difficulties of relocation are recounted in Eleanor Buckles' novel *Valley of Power* from 1945. Buckles tells of the young northern engineer Rootman Jones, who is employed by the TVA to relocate families in an area that is to be flooded by a new dam. Driving around the countryside interviewing the families in order to negotiate with them, Jones is confronted with all the memories and stories that tie the people to the land. Although Buckles clearly embraces the TVA and its promises, she is also sympathetic to the people who are about to lose their ancestral homes. Thus, as Rootman Jones attempts to convince the remaining families of the individual and communal benefits of electricity and prosperity, he finds it difficult to get beyond issues such as community, family, land, and religion; and although Jones is successful in convincing most families, in the end a few have to be transported away from their homes as the water rises and submerges their lands. Although it remains largely positive towards the federal initiative to transform the region, *Valley of Power* nevertheless conveys a sense of regret.

In 1960 the human consequences of relocation were dramatized once more, this time in the form of Elia Kazan's movie *Wild River*. Essentially a love story between a TVA agent and a young local woman, the movie pits the wonders of progress and enlightenment against the noble values of pre-modernity. The movie revolves around an old woman, Mrs. Garth, who refuses to sell her land although her home is about to be flooded by a new dam. As in Buckles' novel, the attachment to the land and its history proves stronger than any monetary or social incentive that the TVA agents can offer, but Kazan avoids portraying Mrs. Garth in an anachronistic or ridiculous fashion. Rather, as TVA agent Chuck Glover—played by Montgomery Clift—makes clear early in the movie, the old woman's stubbornness is the kind of rugged individualism upon which America is built. Although she cannot be allowed to stand in the way of progress, her willingness to fight for her home is something to be admired and respected.

The film opens with black and white footage of the destructions and tragedies wrecked by the Tennessee River as it floods a small town. A devastated man tells of the death and disappearance of his entire family into the raging river, before a voice-over explains how the TVA was created by Congress to prevent the river from ever flooding again. Having thus placed the Authority in a human context the movie cuts to a small Tennessee town and the arrival of the new and idealistic TVA agent, Chuck Glover. He is there to talk the old matriarch into selling her land, a task that proves to be much more difficult than he initially expects. Already during his first visit to Mrs. Garth's island he is thrown into the river by one of her sons after suggesting that perhaps the old woman is senile and unable to comprehend the consequences of her refusal to leave. When he comes back the next day, he witnesses the old woman seemingly trying to make an old black tenant sell his dog to her. The man refuses to sell the dog at any price, and as the situation develops it becomes clear to the viewer that Mrs. Garth, aware of Glover's presence, is actually creating a parable, imitating the TVA's attempts to force her to sell her land. As she subsequently shows Glover around the small island, she further tells him of its history and how her deceased husband worked hard to drain the fields and cut down the trees in order to grow a crop and make a home. She also shows him the family graveyard where her tombstone is already set next to her husband's grave. During their talk Glover slowly begins to understand the old woman, and his sympathy for her cause is only strengthened by his encounter with the young widow Carol, who is the old woman's granddaughter. Carol, however, actually turns out to have some sympathy for the TVA project and understands the necessity for progress, and slowly Glover falls in love with her. Mrs. Garth's pastoral narrative is interesting when seen in relation to the larger TVA narrative of recreating the landscape. Her story after all is one of domesticating a continent, a process that forced away Native American tribes for the "good" of Western civilization. In that sense the two narratives are similar, separated only by time and by the fact that the role of Mrs. Garth and her husband is now being played by a major government agency.

Mrs. Garth's refusal to move and Glover's developing relationship with Carol are far from the only complications that the young TVA agent encounters. When he learns that the workers clearing the river bank are behind schedule, he suggests hiring more men, something that cannot be done without hiring black workers. In spite of being told that using black labor will cause the white workers to quit, Glover nevertheless makes arrangements for the black tenants on Mrs. Garth's land to come and work for him, a ploy he hopes will also convince her to move. This decision pits him against a group of racist businessmen who want him to keep the black workers in a separate work gang and to pay them less than the white workers. Glover agrees to keep the black workers separate from the rest, but insists that the federal government cannot engage in racist practices by paying black workers less than white workers for the same job, a decision that gets him beaten up on more than one occasion.

In the end Mrs. Garth is evicted from her land and moved into a new and modern house on the side of a newly paved road. Sitting on the porch watching the cars pass by she becomes a tragic character that is unable to fit into the new and electrified world of the post-TVA Tennessee Valley. Within days of having been relocated she dies, but as Carol tells Glover, who feels partly responsible for the old woman's fate, there was really nothing else he could have done. *Wild River* balances on an interesting edge, simultaneously celebrating the progress that is to lead the South away from a racist and backwards past, while at the same time lamenting the tragic destiny of the old woman and the noble and stoic values that she represents.

Kazan's movie also shows that no matter how hard the TVA attempted to integrate past practices with new technologies and lifestyles, it was unavoidable that the agency's methods and goals would conflict with the traditions of the region. As some critics have pointed out, by virtue of its wide range of programs the TVA was dangerously close to becoming a federal sociological experiment designed to impose new and improved values on a population rooted in tradition. Such interpretations hardly bothered Arthur Morgan whose mission was decidedly philanthropic, but historians and intellectuals from the region saw a host of problems connected to such an effort. Not only did they see it as a kind of regionalized nation-building, but the TVA's efforts also patronized the cultural heritage of a region that was still having a hard time letting go of the perceived injustices of the Civil War and Reconstruction.

Among the most outspoken of the critics was Donald Davidson, who saw the TVA as "an irresponsible projection of a planned, functional society into the midst of one of the most thoroughly democratic parts of the United States" (Davidson, "That This Nation" 164). Davidson was a part of the Southern Agrarians, a group of writers and poets who, in response to columnist H. L. Mencken's famous critique of the South as a "Sahara of the Bozart," advocated a return to an agrarian society and criticized industrial America for destroying traditions and values important not only to the region but to America as a whole. Nevertheless, the agrarians generally supported the TVA, which they believed could help bring Southerners back onto the farms. But they also feared the loss of regional identity as a result of the new ways of life that the Authority promoted. People who had traditionally pitched horseshoes as a social activity were now being encouraged to sit at home listening to the radio instead, and inhabitants of the valley were indirectly forced to trade in what many felt were important cultural expressions for modern TVA-sponsored temptations.

It is revealing that so much of the criticism directed at the TVA came from intellectuals. College professors, journalists, and artists possessed both the ability and the means to communicate their critique to a broader public, but many of them also remained socially and spatially removed from the problems, and thus were largely unaffected by the New Deal programs. Consequently, to some extent, their criticism was formed by a lack of the immediacy that many ordinary Southerners felt as they were struggling to get through the Depression. Davidson's comment, for example, that the TVA was just another Yankee raid into the South, was both politically and historically insightful, as well as a contextually relevant observation on a part of the Authority's nature, but it was probably one that must have seemed academic and somewhat aloof to most Tennessee Valley farmers whose main concern was to provide for their families.

However, such factors did not necessarily make the criticism any less relevant from a historical and analytical perspective; in fact one of the most poignant studies of the conflict between the TVA economy and a tradition-bound rural South was Madison Jones' 1963 novel *A Buried Land*, which revolves around the hopes for a utopian future and the death of traditional values in the face of change. It depicts the return of a young lawyer, Percy Youngblood, to his hometown Warrington after eight years of absence. The name Youngblood symbolizes the new generation that is about to replace the old, a reading that is further supported by the fact that Percy has chosen to distance himself from his family after an argument with his father about the blessings of the TVA. The father is simply unwilling to accept the kind of utopian scenario that Percy describes to his friend Jesse at the start of the novel: "[In] a few years … you won't know this place. It hasn't even got started booming yet. Right here on the lake, with all that power just a few miles away. There'll be factories" (Jones, *A Buried Land* 5). The promise of electricity, jobs and flood control means less to Percy's father than his home, and so he remains unwilling to leave a farm that is about to be inundated by the waters of the new reservoir. The fight with his father compels Percy to move to a boarding house in town, and later he even refuses to come home for the ceremony when the family graveyard is relocated. His rebellion against his father signifies not merely a break with his own past, but a refusal to honor the past at all.

Percy's clash with the past, however, revolves not around the dispute with his family, but around a secret that he carries with him; a secret about his involvement in the disappearance of a young local girl, Cora Kinkaid. As opposed to Percy, Cora is very much a product of the old ways, and she is first introduced to the story when he picks her up after a church meeting. When she reveals that she is pregnant with Percy's child, he hopes to avoid taking responsibility and convinces her to have an illegal abortion performed. When the operation goes badly and Cora dies from blood loss, Percy and his friend Jesse bury her body in one of the empty graves that is about to be flooded by the new TVA reservoir. Percy and Cora's relationship symbolizes the corruption of the old values, and Cora's blind love for Percy matches Percy's own blind infatuation with the TVA sponsored utopia that he associates with technological and social progress.

Yet upon his return, Percy finds out that the past is not easily forgotten. Not only does a drought threaten to expose the area where Cora lies buried, but her brother, Fowler Kinkaid, has remained determined to find out what happened to his sister. Fowler's presence gradually drives Percy into a guilt-induced paranoia, and when Jesse is murdered, Percy believes that Fowler is coming for him next.

Panicking, Percy runs Fowler over with his car and kills him. Caught in a web of guilt and lies, he attempts to justify his own past by defending a young boy who has killed his father. But when Percy is unable even to convince himself of the boy's innocence, he is ultimately forced to face his own sins. Taking his mother out to the place where he buried Cora, he confesses to his involvement in the young girl's death.

The novel is full of examples of the tension that is created between the past and the present in the post-TVA Tennessee Valley: the death of the innocent Cora, of Fowler, who was both unwilling and unable to let the past go, and Fowler's murder of his own dog, symbolically named Old True, all signify the death of the traditions and values upon which Southern society was built. Warrington is no longer the sleepy ghost town that Percy left behind to go to war, but a changed place due to the local phosphate plant and the new factories that have been attracted by the cheap government power. Old residences, symbolic of the town's past, have been torn down to make way for new shops and businesses. At night the city is lit up by the new street lights, traffic has increased greatly, signalling a changed economic reality, and motorboats seem to be everywhere. A direct consequence of the new economy as well as of the new reservoir's recreational potential, the boats become signifiers of the change that has been imposed upon the region. During a church meeting the sermon is interrupted by the sound of a motorboat on the lake, signalling the gradual replacement of God with electricity and recreation. "He [Percy] was near a window and, later, he could hear from down on the lake the keen wasplike whine of a motor boat. It continued a long time, dividing his consciousness, so that at last the voice from the pulpit reached him like intermittent chilling waves that overflowed his mind. In the end he had entirely lost the thread of meaning in the amiably sunburned preacher's sermon" (Jones, *A Buried Land* 105). The preacher's sunburned appearance seems to suggest that he has also become ensnared by the possibilities of the new recreational landscape, signifying the death of religion in the face of the new consumer society. Immediately after church Jesse invites Percy to go cruising on the lake in Jesse's new motorboat. As they gather speed on the lake they nearly run over an old angler in a flatboat. The clash of old and new technologies in this scene—perhaps a reversal of Huck Finn's encounter with the steamboat, the *Walter*

Scott—is a far less idyllic narrative than the one told by the TVA's display of an antique mill near the new hydroelectric turbines at Norris. In Madison Jones' post-TVA South there is seemingly no place for the values represented by either church or the old fisherman on the flatboat. *A Buried Land* becomes a powerful indictment of the TVA and the modernity it brings to the region.

But it was not only the families whose homes had been flooded that were dispossessed as a result of the TVA's new landscape. The large communities of houseboat dwellers that were living on the Tennessee River also felt the direct consequences of the new planned environment as the river became subject to new technologies and new rules. Most of the houseboating families had moved to the river to enjoy the personal freedom it brought them, while others moved onto the rivers during the Depression in order to seek an economic alternative to unemployment and breadlines. In fact, by 1938 Southern rivers were home to an estimated 30,000 houseboats (Sayre 10). But as the river became increasingly industrial and recreational, the houseboating culture clashed with other interests. On the new river there was little tolerance for either the homes or the nets of subsistence fishermen, who competed with commercial and recreational fishermen for space and who blocked the river for the new and faster boats. Such uses of the river challenged the Authority's control, and for many years government officials did what they could to get rid of the houseboating communities. Few families were stubborn and resourceful enough to fight the TVA for the right to stay, but even for those who did, the Tennessee was a changed river. Not only were the communities dispersed as many followed TVA directions and moved, but the freedom on the river that initially had lured so many away from city and farm life was greatly limited by the many locks and dams. Furthermore, motorboats and waterskiers made it difficult for the river dwellers to find the quiet refuge from society that the old river had offered.

Among the families that stayed on the river were the Sayres.[3] Like so many other young men, Archie Sayre and his brother Ted had been raised to take over the family farm. Their land bordered the river, and Archie's father had allowed a houseboating family to tie up their

---

[3] The Sayres' story is recounted in the introduction to: Maggie Lee Sayre, *'Deaf Maggie Lee Sayre': Photographs of a River Life*, ed. Tom Rankin.

boat on his land. His encounter with the river dwellers made Archie increasingly fascinated by the river culture, and in time he married into the houseboating family. Archie soon came to cherish the independence of the river. He built two different houseboats, a two-room houseboat for the newlyweds sometime in the 1910s, and by the early 1920s a larger three-room boat with room for the couple's two children, Myrtle and Maggie Lee.

Before long Archie and May had become part of the traditional river culture with its individualism, freedom of movement, and subsistence living. During a normal week, Archie would get up early in the morning to put his nets in the river while May often fished with a pole from the houseboat's porch. Two days later Archie would go out in his motorboat to raise the nets and bring home a catch that typically consisted of catfish, buffalo fish, or carp.

Over the next approximately fifty years the Sayres caught and farmed their own food, trading or selling part of their catch on local markets, and supplemented their income by making whiskey. Being subsistence farmers they knew hunger and a singular diet all too well, and consequently they also understood the importance of protecting good fishing waters. But keeping one's spot on the river was a job that became more difficult as traffic on the river increased. More than once Archie had to show his shotgun to chase off commercial fishermen who trespassed onto what he believed to be his fishing grounds, which earned him a reputation for being a hard working, and an occasionally violent, fisherman. At other times the family simply had to move when mussel boats tore up their nets. By all accounts it seems the life that the Sayre family had chosen for themselves was a difficult one. But the ties to the river were strong and the family stayed on the Tennessee and its tributaries until 1971.

Not long after Archie and May had moved onto the river, May gave birth to two daughters. Myrtle was born in 1919 and Maggie Lee the following year. Both daughters were born deaf, possibly as a result of malnourishment during the pregnancies. Being unable to speak and hear, the two girls were isolated by their handicap as well as by their life on the river. Before long, however, they would come to rely on each other for companionship and communication, sharing every aspect of their childhood with each other.

In 1927 the two girls were sent to a school for deaf children in Danville, Kentucky. A local woman had heard of the children's inability to communicate, and persuaded Archie to send them away to school. For the next years, Maggie and Myrtle stayed at the school for nine months out of the year, coming home only for three months during the summer. At the school the two girls were introduced not only to other deaf children, gaining more friends than they had ever had before, but they were also taught how to read and write, and were schooled in mathematics, home economics, and sign language.

When Myrtle died of an unknown illness in 1936, Maggie began feeling increasingly isolated. She went to school for an additional four years until she was 19 years old, but although she was now able to both read and write, and could communicate with her parents as well as with visitors who came to the houseboat through notes and signs, there was no one with whom she shared the same intimate understanding and companionship that she had shared with her sister.

Six years earlier Myrtle had received a camera as a promotional gift from the Eastman Kodak Company who gave away cameras to all twelve-year-old children in order to celebrate the company's 50th anniversary. After her untimely death the camera became Maggie's, and as time passed the photographs she took became a significant part of her communication with the people that surrounded her. Through the images she could relate her experiences and, although she could not have known it then, in time the pictures would serve as documents not only of her own existence on the river but of a vanishing lifestyle and culture.

A collection of Maggie's photographs was published by the University Press of Mississippi in 1995 with the title *`Deaf Maggie Lee Sayre': Photographs of a River Life*. The images present a fascinating account of the river culture of which the Sayres were a part, showing her family and friends, the Sayre houseboat, the daily chores as they are being performed, and the fish they caught. The pictures document her life on the river. Posing for the camera with large catfish they had just pulled out of the water, or taking pictures of her father in the process of selling a fish at the market, Maggie's photographs serve as a visual record of experiences that she would otherwise have been unable to share. But they also function as a window to a culture that was drastically changed as American rivers became increasingly commercialized. Places where the river dwellers could stay became scarce, and whereas houseboating had formerly been a subsistence lifestyle, the new recreational landscape encouraged and promoted a new form of leisure-time houseboating for people who wanted to escape the winter months by sailing down South.

The ranks of the dispossessed did not just include the relocated families and the river communities. As Nancy Grant shows in *TVA and Black Americans*, African Americans were discriminated against when seeking TVA employment, TVA housing, and even when they attempted to use the new recreational facilities. The many difficulties experienced by Southern blacks in relation to the TVA constitute one of the most damaging counter narratives to the ideology of social planning and environmental balance. It is interesting, although not surprising, that the TVA should be mired in this kind of controversy. After all, many of the New Deal programs suffered similar criticism at the hands of civil rights organizations. Yet, the TVA's visionary character and its reliance on sections 22 and 23 of the TVA Act, which gave the Authority the power to plan for the economic and social well-being of the people living in the valley, did seem to promise more than the agency was able to deliver.

Part of the explanation for the TVA's lack of progressivism in dealing with racial issues rests with Arthur Morgan. He was the director in charge of most of the agency's social planning efforts, but he was also known for his elitist beliefs that it was up to the well educated few to ensure progress and to plan for the uneducated many. Morgan openly advocated Booker T. Washington's philosophy that blacks should focus their attention on vocational training rather than on attaining immediate political and educational equality, and he was a firm believer in social Darwinist theories. During his time at Antioch College, Morgan promoted the idea of eugenics, a theory popular in the 1920s that proposed a program of controlled selective human breeding. Such programs often included the segregation and sterilization of groups deemed unfit to carry forward American society, just as they advocated the stimulation of birth rates among the well educated (Talbert 58).

Morgan's fascination with achieving human perfection through social planning remained with him throughout his life. However, already by the mid-1930s he was becoming more moderate, moving away from the idea of segregating the weak and becoming increasingly willing to believe in the possibility of social mobility. In fact, while he was still functioning as chairman of the TVA board, Morgan declined an offer to be named honorary president of the American Eugenics Society (Talbert 15). Nevertheless, it seems plausible that, due to the elitist nature of his core beliefs, Morgan's visions for the Tennessee Valley never seriously included desegregation or racial equality to the extent that African American interest groups and others had hoped for. Rather, when it came to the role of African Americans in the Tennessee Valley, the TVA's, and Morgan's, efforts may best be described, in the words of Nancy L. Grant, as planning for status quo.

Morgan's scepticism regarding racial equality was far from the only consideration guiding the agency in its problematic racial dealings. The South in the 1930s was a difficult place to be advocating civil rights issues, and the TVA board had to consider the political and cultural framework in which it was operating. Neither Harcourt Morgan nor David Lilienthal were eager to formulate a progressive policy on racial matters. Harcourt Morgan, having lived in the South for years, seemed to have accepted the flaws in the system, and although Lilienthal was a liberal his belief in grassroots democracy made it hard for him to ignore the opinions of white Southerners, even when he disagreed with them. As his journal shows, Lilienthal saw segregation and the strained relationship between blacks and whites as a historical and social fact that had to be recognized and taken into consideration (Lilienthal, *The Journals* 516). Change, he felt, would have to come gradually through the education of both black and white Americans and not as something forced upon the people by a government agency.

Nevertheless, the black communities in the Tennessee Valley had high hopes for the TVA, believing that it would become an all-round economic and social boost. Black workers all over the country had been hit hard by the Depression, and in the Jim Crow South their situation was particularly difficult. The decrease in job opportunities meant that poor whites had begun taking over traditional black jobs such as waiting on tables, or working as bellmen and truck drivers. This development forced African Americans into unemployment or onto small plots of land as tenant farmers.

It was clear from the beginning that the TVA could not avoid the issue of race. In an attempt to avoid further punishing the black community the TVA's "Employee Relationship Policy" from 1934 specifically noted that workers should be hired on the basis of skill alone; a public declaration was even made that there would be no differentiation in work or pay on the basis of race or gender (Grant, *TVA* 19-20). Yet at the same time such official statements were made, plans for a quota system were made in order to ensure that African Americans were employed only in numbers equal to their demographic share in the surrounding area. Of course, the quota system could be seen to protect white jobs as well as to secure black employment, a duality that shone through in a 1934 speech that Morgan gave at the black Knoxville College: "We hope that the Tennessee Valley Authority may bring opportunity to the people of this region and, through them, to all Americans. We hope that the black people of this region may have their fair share of this opportunity. … We are trying to employ black people in about the same proportion that they bear to the population. In the territory from which men are employed for the Norris Dam, about six percent of the population is black. We are endeavouring to employ about the same percentage of black labor." Careful not to promise black employment on account of white jobs, Morgan went on to point out that "black and white workers are paid the same wages for the same work", and that in "ways such as these, we are trying to give a fair deal to both white and black workmen" (Morgan 74).

To fully understand the quota system, it must be seen in the context of the political climate of the South in the 1930s. The TVA board was already fighting significant political opposition in Congress due to the controversial nature of the Authority, and the agency was naturally wary of alienating supporters, many of whom were Southern politicians and businessmen. It was difficult to dismiss arguments claiming that equal pay for equal work threatened an economy that had been built around a wage differential for black workers, and that equal wages would lead to increased unemployment for African Americans since most employers preferred white employees. As a result the gap widened between the agency's utopian rhetoric and its practices.

The reality was that it was often difficult for black workers to get TVA jobs. They were frequently denied the application forms which the agency had issued to post offices across the valley, and if they got hold of a form they were discouraged by stories emphasising how the TVA was meant to benefit the white community, or that by signing up they had to work a four-year contract for the government. The TVA did little to get the forms or the correct information out to the black communities, and when black workers were hired it was mostly for unskilled jobs, such as reservoir clearing, and often only on short-term contracts. The high unemployment rates meant that the racial quota was filled quickly, and the TVA did not have to even consider black applicants for skilled positions. A few situations, however, testify to an underlying structure of discrimination, as when in 1938 the Authority hired Francis Steele for a job as an electrician, believing him to be white. When it turned out that Steele was black, he was fired at once, demonstrating the agency's de facto policy of hiring black workers only for unskilled positions (Grant, *TVA* 23).

Often the TVA did not even live up to its own employment quota. At Norris Dam, for example, black workers made up as little as between 1.8 and 2.7 percent of the TVA work force, in spite of the fact that 4.5 percent of the local population in the fourteen counties surrounding Norris was black (Grant, *TVA* 48). In fact, the few black workers employed at Norris were not even working on the dam, but on the margins of the dam area constructing roads. The most frequent excuses when faced with such statistics were that it was too costly to create separate facilities when there were so few black workers available, and that there was a safety risk involved in employing them. Also, because the TVA did not operate with racially mixed work crews, which ironically had otherwise been somewhat common in the Tennessee Valley, often black workers were only hired if there were enough to form separate gangs.

Although officially refusing to pay black workers less for doing equal work, white and black workers were rarely paid the same wages. One way around the rules, as Nancy Grant shows, was to classify black and white workers differently (Grant, *TVA* 50). The agency also adjusted to the racial institutions of the South in a number of other ways: separate toilet and drinking facilities were installed at all the TVA's dams, and the technical assistance, expertise, and financial aid that were made available for agricultural programs were distributed through existing channels, which meant that only white agricultural colleges benefited from it. Also, the selection of demonstration farms happened through local channels, resulting in only white farmers being chosen for the program which included the free use of new technologies, free fertilizer, and a reliance on government expertise.

Furthermore, since a large part of the agricultural system in the region was built on an owner-tenant relationship, in which most owners were white and many tenants black, it was hoped that once the TVA began buying land for their reservoirs compensation would be paid directly to the tenants. But since most landowners were interested in perpetuating the economic and social structure of the owner-tenant relationship, and because large-scale farmers held political sway, ultimately compensations were paid directly to the owners who would then distribute the money. This of course helped sustain a situation in which tenant farmers remained dependent upon landowners.

The Norris Dam projects did not just set a poor example with regard to employment statistics, but among the most visible of the many discriminations against African Americans committed by the TVA was the refusal to let black workers live in the planned community of Norris. The official explanation was that by Tennessee law the TVA should then also provide separate facilities, including schools, which would raise prices considerably. As early as 1934 and 1935, articles began appearing in black publications criticizing the TVA for its discriminatory practices, and Norris in particular received much attention because the town was so often praised in the media as an ideal community. Cranston Clayton regretted that "the Negro … [was] to be absolutely excluded" from the much needed job training programs offered at Norris, and John P. Davis saw Norris as yet another example of the "lily-white reconstruction" from an administration that had once again failed to bring black Americans civil rights (Clayton 111; Davis, "A Survey of" 11).

African Americans were also prohibited from using the new recreational facilities that had been created around the dam. Although the parks at Norris had never been declared segregated spaces, black visitors could not rent cabins, boats, or even use the restrooms. This was an ironic turn of events considering that Knoxville was far less segregated than other parts of the South. Black Knoxvillians had a history of voting in significant numbers in elections at all levels, and they could patronize most businesses and restaurants. Although most African Americans still occupied traditional positions as domestic servants or performed other kinds of unskilled labor, there existed a significant black middle class, educated at the local black Knoxville College, both willing and capable of taking advantage of the recreational offers at Norris. In a very real sense the TVA actively created a segregated social structure, racializing the new middle landscape. If Norris was a demonstration of how the American future could look if only planned correctly, the practice of racial exclusion assumed an almost entirely white and segregated future society.

In response to the exclusion of African Americans from Norris and the lack of facilities in the parks around Norris Dam, political pressure for the creation of a separate recreational space for black people mounted. Part of the idea behind the recreational areas had been to allow farmers, whose farms had been diminished by the reservoirs, to supplement their income by running concession stands or renting out boats or fishing gear to visitors. By denying African Americans the right to use the recreational landscape, the TVA effectively excluded them from participating in the new tourist economy. By 1937 the possibility of a "Negro Park" was being discussed internally in the TVA, and on December 6 of that year it was decided to form a three-man committee with representatives from the Tennessee State Planning Commission, the Tennessee State Department of Conservation, and the TVA (Draper, "Negro Park"). Within two months the committee's report was ready, concluding that the best place to locate a park for African Americans would be at Norris Lake. The committee argued that much of the ill feeling towards the TVA which African Americans all across the nation bore on account of the exclusion and segregation practices at Norris Dam and town, might be repaired if a park was established in that racially contested area. Also, the report noted, Norris was a better location than Chattanooga, which had also been suggested, because the black population of Knoxville had more money to spend and thereby seemed more likely to take advantage of the recreational opportunities. Furthermore, the attraction of the Great Smoky Mountains National Park would appeal to many black tourists from all over the country ("Report of the Committee").

The report was commented on by people from the TVA's Legal Department, the Reservoir Property Management Department, the Budget Office, the Department of Information, the Personnel Department, the Knoxville Negro Committee, and an African American representative from the Department of the Interior. These opinions were considered by the Department of Regional Planning Studies which then prepared a summary of the findings for the TVA board by January 20, 1938 (Draper, "Negro Park"). In the summary, pros and cons of the proposed projects were considered, including the size and financial status of the black populations in Knoxville and Chattanooga. The fact that African American groups, some counting as many as 150 people, were frequenting the park in spite of the lack of facilities seemed to work in favor of a location at Norris (Tennessee Valley Authority, "Review of Report"). Not only did it show that people were interested in the location, but also a separate park at Norris might make it easier to deny African Americans access to the other recreational areas created there.

Yet in spite of the committee's report and the positive summary, not many within the TVA favored Norris as a location for such a park. In an internal report from July 1938 the Norris area was deemed unsuitable "for the usual cooperative activity to proceed in relation to a Negro park development" (Howes i). It made sense, according to the report, to prioritize the needs of the underprivileged as opposed to the wealthier black Knoxville community, and also the black population in Chattanooga and the surrounding area was several times larger than the one in the immediate Knoxville area. As an alternative, the report suggested, land at Norris could be leased to black organizations that would then be responsible for its development and care, as long as it fulfilled TVA guidelines.

As the unpublished reports show, the idea of locating a park for African Americans at Norris was not seriously considered by the TVA. Norris was already an established fact, an existing landscape infused with ideology and vision, and there was little desire to alter that creation by including those who, many felt, did not belong. Although no official decision was made regarding a park at Norris until 1938, work on the appropriately named Booker T. Washington Park, located on the shores of Chickamauga Reservoir in Chattanooga, had already started in 1937 and was finished a year later. The Chattanooga park, which was created in cooperation with the CCC and WPA on land which the TVA had leased to the state, certainly made it easier for TVA officials to explain why a park for African Americans was not being built at Norris.

Booker T. Washington Park was a disappointing effort to extend recreational opportunities to African Americans. Nearly a decade passed before the park was even operational. The swimming pool was kept closed for years due to a lack of pumping equipment, and there were no restrooms in the park. Other plans for facilities were postponed when America entered World War II, and it was not until after the war that the park became fully operational. Furthermore, the park's location was within sight of an existing industrial site which diminished its scenic value greatly (Grant, *TVA* 88).

Of course, there was no reason that parks could not have been built at both Norris and Chickamauga, but it seems clear that the establishment of a park for African Americans in Chattanooga was an easy way to alleviate some of the criticism directed at the agency. In reality, the Booker T. Washington Park was never given any of the same social and civic character as the agency's other recreational projects. Rather, the existence of a park for African Americans was carefully measured against its economic feasibility.

Studying the TVA's approach to racial issues might give the impression that there are few redeeming aspects of the agency's attitudes towards the black population of the Tennessee Valley. Yet, as J. Max Bond, the organizer of the TVA's black construction workers in the mid-1930s explained in an article, black workers who had been used to earning between 37 cents and 1 dollar for an entire 12-hour work day now earned between 45 cents and 1 dollar per hour for a work day of only six hours (Bond 146). Even though the TVA did adopt many of the discriminatory practices of the region, the agency was also bettering the economic standing of many Southern blacks, even if they were never truly a part of the overall vision for the region.

Although the TVA was rooted in a Depression-age mentality and a desire to transform the region economically, underlying many of the agency's early efforts was an ideology of environmental balance and conservation, an ideology that contained a critique of modern society for having distanced itself from the environment it relied upon for resources as well as for inspiration and escape. The answer to this critique seemed to be a pastoral element that allowed the TVA's planners to incorporate a particular vision of the American past into its plans for the future; in a sense the TVA engaged in a recovery narrative, reclaiming a ruined landscape and transforming it into a fertile, sustainable, and user-friendly environment. Traditional agriculture was converted into dairy farming, fields became pastures, and cash crops like tobacco and cotton gave way to the cultivation of fruits, nuts, and berries. Programs promoting forestry communities saw industrial workers relocated to the woods where they were given homes on small plots of land and taught how to manage the forest. In the demonstration community of Norris, architecture and nature blended, just as the town was designed to enter into a dialogue with the surrounding landscape. Provisions were even made to make it easier for residents to supply incomes with part-time farming. And a number of recreational programs attempted to re-establish man as a part of the landscape through escapism and play. It is of course only one narrative of many to which the TVA lends itself, but the agency was an important factor in the development of the pastoralized middle landscape. Having moved past the unintended pastoral space, the Authority's planners relied in part on simulations of the pastoral in order to create what they believed to be a better future.

Four

A Breakdown of Ideologies: The Tellico Dam and the Tennessee-Tombigbee Waterway

I

The controversies surrounding the construction of large public works projects such as the Tellico Dam in eastern Tennessee and the Tennessee-Tombigbee Waterway—also known as the Tenn-Tom—in northeastern Mississippi and western Alabama, are often ascribed to the changing public attitude towards large-scale federal developments, and rightly so. By the 1960s the public belief in the benefits offered by such undertakings was rapidly diminishing. Both Tellico and the Tenn-Tom might have been celebrated had they been constructed in the 1930s, but by the mid-1960s they were regarded with skepticism as environmentally arrogant and unnecessary pork barrel projects, which made them easy targets of the growing political grassroots activism. But the public attitude towards the Tellico Dam and the Tenn-Tom Waterway was more than a result of the political climate of the times. It was also partly due to a failure to root such projects in an ideology that might have engaged the American imagination. After a decade of conservative politics and general affluence, it was difficult for many to relate to projects with little to offer except increased industrialization. Faced with such criticism, the Tennessee Valley Authority and the US Army Corps of Engineers both sought to infuse their projects with a sense of the pastoral in order to improve public relations and failing benefit-cost ratios. By promoting issues such as environmental balance and recreation they hoped to soften the projects' images, but such efforts also emphasized the breakdown of the pastoral as ideology.

In spite of the TVA's attempt to recreate at Tellico the comprehensive approach to planning that had characterized the Authority during its first decade, the project seemed to contradict much of what the agency had worked to build in the 1930s. The history of the

Tellico dam suggests that multi-purpose planning was undertaken only to the extent that it was useful in gaining Congressional approval, and the resource conservation efforts that had been so important to earlier TVA projects were all but non-existent. Unlike the Tellico Dam, the Tenn-Tom was a purely economic venture from the start, and the Army Corps of Engineers only incorporated a pastoral ideology later as a response to criticism of environmental and economic irresponsibility. At both Tellico and the Tenn-Tom the narrative of retreating back into nature was no longer imbued with a vision of nature/culture reintegration, but merely functioned to sell problematic projects to Congress and the public. From having been a utopian and visionary principle of public planning, projects like the Tellico Dam and the Tenn-Tom accentuated the middle landscape's development towards a new phase, the hyper-pastoral, in which the pastoral design was released from the meanings with which it had formerly been associated.

Originally proposed in the mid-1930s as the Fort Loudon Extension project, the Tellico Dam was re-launched in 1959 as the answer to the TVA's post-war identity crisis; a symbol of the Authority's attempt to justify its own continued existence in a time when many felt that it was no longer needed. Breaking with its earlier ideology of comprehensive planning, by the 1950s the TVA had "evolved from a benevolent paternalism into the biggest power producer, biggest strip miner, and single biggest polluter in the United States" (Reisner 326). In that context, any attempt to return to the earlier vision of responsible multi-purpose planning would necessarily prove a problematic gesture.

In some ways it was a development that had already begun when Arthur Morgan was forced to leave the board of directors. Lilienthal, who was becoming increasingly influential within the agency, had always been against the kind of top-down planning that Arthur Morgan represented, and in Morgan's absence he successfully shifted the TVA's emphasis away from the earlier multi-purpose developments to the narrower goals of power production, navigation, and flood control.

The war also propelled this change. In 1941 the newly established Office of Production Management (OPM) had requested an additional 100,000 kilowatts to be used primarily for the production of aluminum to build more bombers as part of the defense crisis and the lend-lease agreement with Britain. The immediacy of the war effort meant that the goal had to be reached by the winter of 1943, and demands for power production did not diminish when America entered the war. Suddenly, the TVA also had to deliver additional millions of kilowatts to a project known only as the Clinton Engineer Works, a cover for the top-secret Oak Ridge nuclear facility that helped build the world's first atomic bomb.

Although another major dam was being built at Fontana at the time, the only planned project that could match the OPM demands in terms of timeframe and production capacity was the Douglas Dam. Initially, however, the Douglas Dam project fared poorly. Harcourt Morgan regretted the loss of between ten and fifteen thousand acres of fertile farmland and attempted to stall construction, and eventually he was joined by Tennessee Senator Kenneth McKellar who, backed by local farmers, also protested the plans about which he felt he had not been properly informed (Hargrove 56-57). But the surprise attack on Pearl Harbor silenced all protests. The dam was constructed in an impressive 12 months and 16 days, yet Douglas Dam was more than an example of wartime efficiency. It signaled a victory of engineering over politics, and a prioritizing of technical aspects over the smaller programs such as land planning, reforestation, and recreation. It was the start of a new era for the TVA.

The war effort and Lilienthal's leadership began the TVA on a new course leading away from its original multi-purpose developments. The booming economy and the political climate of the post-war years meant that there was little incentive or opportunity to return to regional planning. Even when a director inclined towards comprehensive planning was appointed to the board—as was the case with Gordon Clapp in 1946—the TVA still depended on Congress, and the Republican victory that year meant that there was little willingness to approve anything that even resembled the New Deal. The subsequent election of Eisenhower and the general conservatism of the 1950s offered no opportunities for Clapp's TVA to get back to its multi-purpose planning roots.

The increased demand for electricity in the consumer-driven and suburbanized 1950s created an unprecedented demand for power. But with no obvious locations left on the Tennessee, and lacking political fervor for starting projects on the tributaries, any new power production would have to be non-hydraulic. As a result, in the period between 1949 and 1959 the TVA began construction on eight coal-fired plants, adding three more in later years. To feed the new plants, coal had to be mined, and since surface mining was the least expensive and the fastest, the agency became involved in a strip-mining program that systematically ruined the landscapes of the region it had been trying to rebuild only twenty years earlier. Further emphasizing the shift in values, an increasing amount of the power produced by the TVA went to non-private purposes. As Gordon Clapp writes, by the mid-1950s approximately 50% of TVA generated power went to federal atomic-energy facilities (Clapp 10). Clearly, the people of the Tennessee Valley no longer were the TVA's primary customers.

Being an ordinary power company, however, was not a role that suited the directors of an agency that continued to pride itself on a visionary and comprehensive approach to regional planning. But it was not until early February 1959, when general manager Aubrey Wagner gave his, in TVA circles, famous opening speech at a brainstorming session at the Watts Bar hydroelectric plant that the Authority initiated a serious effort to reclaim what many felt had been lost in the years since the war. In his speech Wagner noted that for years no multi-purpose projects had been proposed and that it was time to reinvent the agency (Wheeler and McDonald 4). Wagner's opening statement marked the beginning of what would prove a twenty-year long battle to build the Tellico Dam.

Wagner had been with the TVA long enough to recognize the authority's heritage in regional planning. Educated as a civil engineer at the University of Wisconsin, he gained employment with the TVA in 1934 working as an engineering aid. It was a job that had him working along the Tennessee River and its many tributaries and which gave him ample opportunity to familiarize himself with the landscape and the people living there. After nine years he became a navigational engineer, and in 1954, having spent three years as assistant manager, he became the agency's general manager. Recognizing his abilities as well as his loyalty, President Kennedy made him a member of the board in 1961 and chairman the following year (Tennessee Valley Authority, "Mr. TVA"). Because of his background, when Aubrey Wagner presented his manifesto at the Watts Bar Dam conference in 1959, people listened.

Wagner believed in integrating Lilienthal's focus on power production with a broader approach to planning, and he saw the development of the Tennessee's tributaries as a logical path for the

TVA to follow. But Wagner's vision for the Tellico area did not incorporate the sense of environmental balance that had characterized the pastoral vision of early TVA projects. Rather, Wagner saw industrialization as the answer to what he felt the region needed (Wheeler and McDonald 31). By industrializing the countryside he believed it was possible to keep young people from moving to the cities. Industry primarily meant jobs and money, and these goals, according to Wagner, could be achieved by damming the Little Tennessee River.

But Wagner was guided by the false assumption that the area was lacking in industrial development. Much of Wagner's own field experience dated from the 1930s and early 1940s, and the original plans for the Tellico project were also from that period when the entire valley region was still suffering from a host of economic and social problems. By the late 1950s, however, the area was far from isolated, and residents saw little need for further federal intervention. The Tellico area had already benefited from the cheap government electricity and, although some communities had remained somewhat isolated, many of those living near the proposed dam had chosen the kind of life they were living, as well as the kind of community they wanted to live it in.

To make matters worse, there were few justifications for building a new dam. Tellico Dam would increase power production only slightly since the dam itself produced no electricity but only stored and diverted the flow of water to the nearby Fort Loudon Dam via an 850 foot long canal. And even then it only increased the production capacity by two percent of the power generated by the agency's new coal plants (Reisner 324). There was some risk of flooding in the area, but the dam only offered a minimum of protection. The proposed reservoir capacity provided seven inches of flood control, far less than what was needed in an area where in 1973 "the St. Patrick's Day flood in Chattanooga crested at 540 inches above flood level" (Ballal 586). And whereas earlier TVA projects had focused on issues such as erosion and bad farming practices, the Little Tennessee Valley was a beautiful and varied landscape that included both flat and fertile bottomland for farming, gently rolling hills, and even areas of wooded wilderness where wild animals could be observed (Wheeler and McDonald 48-49). As Marc Reisner describes, the valley fitted the archetypal pastoral scenario: "With its pretty white clapboard houses and its well-tended little farms, the valley was a beautiful anomaly, a place more at home in the nineteenth century" (324). Adding insult to injury, recreation at Tellico was already thriving. The area was known as one of the best trout fishing environments in the Southeast, and the free-flowing Little Tennessee River attracted thousands of recreational visitors each year. A study done by the TVA and the Tennessee Game and Fish Commission estimated that as many as 24,000 trout fishermen came to the Little Tennessee every year, spending over 80,000 dollars in the process (Wheeler and McDonald 98-102). Apart from trout fishing, visitors also came to the area to enjoy activities such as hiking, bird watching, or to observe other forms of wildlife.

Thus, apart from the desire to attract industry, building a dam on the Little Tennessee River seemed to make little sense. Yet the project went ahead, and in order for it to succeed, the agency's land purchase policy had to be reversed from minimal purchase to overpurchasing. For comprehensive planning purposes more land was needed than what was going to be flooded. What Wagner needed was the kind of land purchase policy for which Arthur Morgan had fought so hard in the early years. But even as the reversal of the land purchase policy became a reality, a study of the agency's actual shoreline property holdings shocked Wagner as well as the board of directors. Since the early 1940s much of the valuable real estate that should have ensured proper shoreline developments had been sold to, and developed by, private entrepreneurs. The board readily admitted that too much land had been sold and immediately put a stop to all future sales (Wheeler and McDonald 19-20).

The reversal of the land purchase policy as well as the decision to stop all sales of shoreline properties sent a clear signal to Wagner that he could proceed with his multi-purpose project. But he still needed a benefit-cost ratio that would convince the TVA board, Congress, and the project's potential critics that it was economically sound. Thus, shortly after the Watts Bar Dam conference Wagner created the Future Dams and Reservoirs Committee (FUDAR) to discover ways in which to improve Tellico's benefit-cost ratio. The FUDAR report was essential in convincing the TVA board that Tellico was a sensible project, because traditional economic benefits such as power production, navigation, flood control, and even the development of the reservoir shoreline only balanced the budget at best. Taken together, flood control and navigation constituted a little less than half of the project's total benefits, and even when a third and fourth generating unit were added to Fort Loudon Dam, the amount of hydroelectric power generated constituted a mere 18% of the estimated project benefits (Elliot, "The Tellico Project" 41). A benefit-cost ratio that only just balanced would not convince the board to green-light the project, and would certainly not sway Congressional appropriations, especially since damming the Tellico area would confiscate land where an estimated 28 million dollars worth of agricultural products were produced every year. Consequently, FUDAR was to come up with ways of quantifying new benefits. It was a task that proved almost impossible, and by 1960 the committee had come up with nothing that could help Wagner justify the project.

The three areas that FUDAR eventually did identify as possible benefits were recreation, general economic benefits, and the highly controversial land enhancement, which meant a scheme of buying land only to develop it and sell it at a profit. A 1963 report prepared by the Project Planning Branch arrived at a 1.5-1.0 benefit-cost ratio, a figure based in part on land sale profits estimated at 10,500,000 dollars, all of which would be returned to the Treasury in order to reduce federal costs (Elliot, Untitled; Elliot, "The Tellico Project" iii). But the plan was controversial because it meant that the TVA would enter the real estate market, and considering how private power businesses had reacted to the agency in the 1930s, that plan was likely to raise a storm of protests. Also, the idea of including land enhancement as a benefit was tricky since the rise in property values adjacent to the new reservoir partly overlapped with recreational developments. Furthermore, many, including Wagner, were worried that buying land only to resell it at a profit might prove illegal for a government agency. Even if that proved not to be the case, methods for calculating the possible land enhancements were too unreliable as the basis for such a decision. Nevertheless, because land enhancement was a necessary part of the project's economic justification, it was included, and additional land was even added to increase potential profits (Wheeler and McDonald 94-95).

As opposed to the earlier TVA projects, the middle landscape ideology was largely absent from the Tellico plans. Recreation was included, but it was no longer part of a vision to reintegrate man and nature. Instead, it became a means to an end that, somewhat ironically, exchanged an unintended pastoral landscape for one that was designed. As the 1963 report recognized, the benefits of recreation were "vital to the achievement of the full economic value from a project" (Elliot, "The Tellico Project" 1). But benefits from recreation remained difficult to quantify, especially since the area was already attracting anglers, hikers, and hunters. It was believed that the reservoir would attract large numbers of bass fishermen, but damming the river would also ruin the existing trout fishing, and it was doubtful to what extent a new reservoir would appeal to a vacationing public in a region that was already full of reservoirs offering similar experiences. After all, the Fort Loudon Reservoir, which was located right next to the proposed Tellico reservoir, also offered bass fishing, boating, and bird watching. In a 1977 letter to the editor in *BioScience*, S.K. Ballal, a Biology Professor at the Tennessee Technical University, noted that it was doubtful what the recreational appeal of the Tellico Reservoir would be in an area where "there are 22 lakes within 50 miles." Especially, he continued, considering that "the Little Tennessee is the last big, clean river in Tennessee, and considered to be the finest trout river in the Southeast" (Ballal).

As Wheeler and McDonald have shown, all of the basic assumptions that made the basis for the dam were either flawed or outright false. It was believed that the area would remain "economically static without the project"; that "all economic progress … should be attributed to the project"; that potential benefits were essentially guaranteed; that the project had no negative impact on the area; and that project costs were under control (92-93). The less than nuanced nature of most of these assumptions was most likely intentional in order to make it easier to portray the project as economically healthy.

But the benefit-cost ratio, vital to gaining Congressional approval, was still far from satisfactory. The pressure was on FUDAR to come up with non-traditional benefits and to discover ways in which to quantify them and include them in their reports. But by 1965 no new ways of boosting the benefit-cost analysis had been found, and in May of that year a new study merely repeated the findings of the 1963 report (Elliot, "The Tellico Project" 41). As in the earlier report, the new study relied on a number of more or less inflated and speculative calculations, and interestingly the so-called General Economic Development was by far the largest post on the balance sheet, accounting for a third of all the benefits attributed to the project. The report reiterated the belief that a total of 22,000 acres of shoreline property would be enhanced by the project and proposed that the agency purchase 16,500 acres of this land before selling it approximately ten years later at an estimated net recovery of 10,500,000 dollars (Elliot, "The Tellico Project" 40-41). Planners were also confident that the reservoir would attract industry, which in turn would create an estimated 6,600 new jobs (Elliot, "The Tellico Project" 37). Yet as time would show, few businesses were to move to the area, firstly because there was no guarantee that they could find a suitable workforce in the predominantly rural area, and secondly because they would have to compete with the industrial giant ALCOA, which already had a plant there. Fishing was included as a major recreational asset, but instead of subtracting the trout fishing that would be diminished or disappear, the report speculated that trout fishing might be continued in spite of the reservoir. It also stated that wildlife populations in the area would not be affected by the reservoir, a somewhat optimistic assessment considering the industrial developments that were expected along the new shorelines (Elliot, "The Tellico Project" 9). As the project progressed, the "tail of justification began to wag the dog" (Wheeler and McDonald 90).

By the early to mid-1960s Tellico had become more than just a project to Wagner. The dam embodied the agency's new mission; it was the answer to the search for a place in post-war America, and it seemed that Wagner wanted to complete the project even if it had no economic justification. So determined had he become that in late 1964—even before the new FUDAR report was officially ready—he sent the Tellico proposal to the Bureau of the Budget to have it approved as an official part of the 1966 budget. Wagner's hurried decision was probably partly based on a strong suspicion that the benefit-cost ratio was not going to improve in the new report, but might also have had to do with the political climate of the time. The election of Kennedy in 1960 had created a situation favorable to the agency's return to multi-purpose planning, and with the democratic landslide election earlier in 1964 and Lyndon B. Johnson's proposed Great Society program, a dam at Tellico seemed closer than ever before. But Wagner did not get all he asked for. Because of the escalating costs of US involvement in Vietnam, the Bureau of the Budget eventually asked the TVA to choose between Tellico and a project known as Tims Ford. Wagner chose Tellico, and the project was approved with almost six million dollars on the 1966 budget (Wheeler and McDonald 105). It would, however, be another year before the TVA received appropriations for Tellico. Having first been defeated in the House Public Works Subcommittee, which shifted the money from Tellico to the Tims Ford project, Wagner subsequently won the support of the Senate, before finally being defeated in the House-Senate conference committee where funding was once again shifted to Tims Ford (Wheeler and McDonald 105-108). That the TVA could not win over the House subcommittee was in part because of Joe L. Evins, a Tennessee Congressman from the area in which the Tims Ford dam was located, but it also had to do with Tellico's notoriously shaky benefit-cost ratio.

Among the best examples of how the TVA's priorities had changed was the debate over the creation of Timberlake, a model city to be located by the new reservoir. First proposed in 1965, the project was originally known as the Tellico New Town but was given the more idyllic name of Timberlake in 1969. It seems only natural that Timberlake would draw comparisons with the town of Norris, but Timberlake lacked the environmental and social ideology that had been so central to the construction of Norris. At Timberlake there was no sense of a fundamental social vision or of tying the community to the surrounding landscape and the natural resources. Construction plans did not involve using local materials or an architectural style that would allow it to blend into the landscape. Instead, some plans called for the construction of modern skyscrapers overlooking the reservoir. The closest it came to employing a pastoral design was through ideas for the beautification of the shoreline, and the closest the project came to a social vision were plans for a number of neighborhood schools designed to evoke "memories of a mythical golden age of youthful innocence, parental control of schools, and walking girls home from school with a stop for a soft drink on the way" (Wheeler and McDonald 170).

Timberlake lacked a clear sense of direction. Every division within the agency had its own vision of what the community should be. Some felt that it should be a corporate city, others wanted it to be a Knoxville suburb, a recreational community, or even a retirement center; and while some wanted it to be for all people regardless of class, others wanted it to be predominantly upper-middle-class. In other words, Timberlake was "a city in search of its own definition" (Wheeler and McDonald 159). But whereas Timberlake lacked a social and an environmental vision, it was based on an economic agenda. Above all else it was thought up to attract industries to the area, and the 6,600 new jobs that had been predicted in the 1965 report, in the new Timberlake estimates suddenly, optimistically, ballooned to 25,000 in order to boost the benefit-cost assessments (Neely).

Politically and economically Timberlake was problematic, and there was little chance of getting Congress to appropriate the construction of an entire town. Thus, the TVA began to look for a corporate developer. By the 1960s the white middle class was moving to the suburbs in increasing numbers, leaving the inner cities to deal with a series of social problems. This migration spurred an interest in new model communities, and before long a number of corporations were looking to profit from the trend. Apart from being a sound investment, planned communities also made it possible to attract workers who wanted to raise their children under more secure and controlled circumstances. When General Electric announced its interest in building a number of planned cities, the TVA immediately contacted the company. It did not take long for General Electric to pass on the TVA's offer, arguing that the lack of a clear source for population and job expansion meant that a venture in the Little Tennessee Valley was too uncertain (Wheeler and McDonald 161). It would not be the last time the TVA received that answer. In the latter part of the 1960s a number of other possible developers also declined the offer to build a planned community at Tellico which implicitly undermined the benefit-cost argument that businesses would come flocking to the area once the dam was built. Not until the early 1970s did the TVA find a possible developer, and for a while the plans to put Boeing in charge of building Timberlake looked as if they might bear fruit. But Boeing also had trouble recognizing the feasibility of the project, and by 1975 the company formally withdrew, leaving the project dead in the water.

Tellico Dam attracted criticism as early as 1961 when it was rumored that the TVA was seriously planning the dam. The early opposition, however, was poorly coordinated, consisting mostly of locals backed by the Tennessee Game and Fish Commission, the Tennessee Outdoor Writers Association, some local industrialists who did not want the competition, a staff member at the Knoxville Chamber of Commerce, and an editor at the *Knoxville Journal* (Wheeler and McDonald 68). Rather than rallying around some larger political or social concern, it seems that the difference between early supporters and opponents came down to their view of community. While some wanted the jobs and economic prosperity that the TVA promised would follow in the wake of the project, others felt that the dam was unnecessary and that it would destroy that which they held most dear about the towns they lived in (Wheeler and McDonald 67).

In the latter half of the 1960s, the debate over the Tellico Dam heated up, becoming national in the process. Criticism increased as more people began realizing that the dam was becoming reality. Many undoubtedly had trouble relating to the abstract nature of the project, until plans for Tellico suddenly became real. Some criticism had to do with the general uncertainty regarding how large an area the TVA would need. As construction on the dam began in 1967 there was still no clear understanding of exactly how much land the agency would take. This created a sense of frustration among land owners who did not know if they were being expropriated. But of the families unwilling to move, few had the resources to stand up to a federal agency, and most simply left when told to.

Tellico's opponents tried to boost their ranks by, somewhat unsuccessfully, engaging the eastern band of the Cherokees in the fight against the TVA, and in 1965 Supreme Court Justice William O. Douglas, a known wilderness advocate who had written several books about the topic, was invited to come and visit the area. Douglas was furthermore famous for his distrust of government, and although his visit was largely symbolic, it did help bring the Tellico issue to the attention of the national media. During his visit Douglas participated in public events such as meeting with a delegation of distressed Cherokees who were concerned about the flooding of old towns and burial sites, and went on a trout fishing trip in order to demonstrate the free-flowing river's scenic and recreational appeal (Wheeler and McDonald 83-85).

But not until the early 1970s did Tellico's opponents manage to unite and coordinate their efforts. Up until that point most of the criticism had centered on attacking the benefit-cost ratio and disproving the reservoir's supposed attraction to industry and re-creationists. But as the environmental movement of the 1960s gained momentum, it united Tellico's opponents behind the new green cause. The new environmentalism was the result of general anti-establishment and anti-corporate sensibilities, with concerns such as the continued fear of nuclear fall-out from the military's atomic tests in the 1950s, a growing number of inner city smog alerts, and Rachel Carson's hugely popular *Silent Spring* from 1962, a book that showed the effects of DDT on an American lake environment (Anderson, "The Movement" 347). Also, the suburban sprawl of the post-war years had brought with it a growing environmental awareness that fuelled the new movement.

One of the main issues with which environmentalists found fault was the TVA's reluctance to file an Environmental Impact Statement for Tellico. Ever since the federal government had passed the National Environmental Policy Act (NEPA) in 1969, it had become policy for federal and state governments to promote environmental quality. No project could proceed without first having established the impact it would have on the environment. Although Wagner argued that there was no need to file one since the project had been approved before NEPA was passed, the lack of an Environmental Impact Statement was hurting Tellico's public image. In an effort to make the problem disappear, the TVA worked on putting a statement together. However, a circulated draft of that statement revealed a number of problems with the dam and was quickly deemed inadequate.

In 1971 the Environmental Defense Fund (EDF), a non-profit group founded only four years earlier, in 1967, opened a lawsuit against the TVA in an attempt to stop the project because an Environmental Impact Statement had not yet been filed. The court ruled that Tellico was subject to NEPA and that the agency therefore had to file an environmental impact statement before continuing work on the dam. The injunction, albeit only partial—work on purchasing land and building new roads could continue—forced the agency to file an Environmental Impact Statement, and although it was completed by February 1972, only one month after the injunction had been issued, construction was eventually halted for 18 months due to a number of problems with the statement.

Work on the dam eventually resumed, but the Authority's troubles were far from over. In August of 1973 the discovery of a small fish, the snail darter, looked as if it might succeed in stopping the dam. The new species was discovered by zoologist David Etnier from the University of Tennessee, who had testified as an expert witness at the EDF trial on behalf of the Tellico opponents. Four months after Etnier's discovery, the Endangered Species Act of 1973 was passed, recognizing that in order to protect endangered species it was also important to preserve the "ecosystems upon which endangered species and threatened species depend." This meant trouble for the TVA, and Tellico's opponents knew it. Work was immediately begun to have the snail darter placed on the endangered species list, a goal that was reached within a year. The snail darter was now officially endangered, and its only proven habitat at the time was a 17-mile stretch of the Little Tennessee River that was about to be dammed.

Meanwhile, as was the case with the Environmental Impact Statement, the TVA once again chose to argue that the Tellico project was from before the Endangered Species Act and therefore was not subject to it (Wheeler and McDonald 192). Opponents did not agree, and when in 1976 the U.S. Fish and Wildlife Service declared the Little Tennessee River a critical habitat they had enough to go to the courts once again. After several attempts to have construction on the dam stopped, by 1977 it looked as though opponents might succeed in spite of the fact that the dam was almost completed by then ("Tellico Dam Work"). That year the Sixth U.S. Circuit Court of Appeals ruled that the TVA was subject to the Endangered Species Act, and that all work on the dam had to be stopped immediately. The agency appealed the verdict, but on June 15 1978 the Supreme Court confirmed the verdict, arguing that the uncompromising terms of the Endangered Species Act allowed no room for the TVA to complete the dam, even if millions of dollars were lost as a result of such a decision ("The Darter"). The only option left to the TVA, if the 140 million dollar dam was to be saved, was to gain a Congressional exemption from the Endangered Species Act.

Although the court ruling went against the TVA's interests, the uncompromising character of the Endangered Species Act on which the court had based its decision resulted in an amendment to the Endangered Species Act that might save the Tellico Dam. The amendment called for the formation of an interagency endangered species review committee that would be able to review cases in which large-scale projects came up against the act. The committee consisted of "representatives from the Secretaries of the Interior, Agriculture, and the Army, in addition to the administrators of the Environmental Protection Agency, and the national Oceanic and Atmospheric Administration, The Chairman of the Council of Economic Advisers, and a representative from the affected state" (Reisner 327). The make-up of the seven-man committee meant that opponents had little hope for a victory.

But once again the TVA's hopes of finishing the dam were in vain. In a unanimous verdict that surprised everyone, the committee ruled in favor of the snail darter. Adding insult to injury, the committee's decision was primarily based on what it felt was a horrible benefit-cost ratio. In fact, as one committee member stated, even though the project was 95% complete at the time, the costs of finishing it were not worth the benefits (Reisner 327).

Then in September of 1979 it was suddenly all over. President Carter signed a bill exempting the Tellico Project from all restrictions in all federal laws, striking a devastating blow against the opponents to the Tellico Dam and to preservationists all over the country ("Tellico Follies"). It was an odd decision for a President who had included the Tellico Dam on his infamous "hit list" of unnecessary water projects, and who had declared war on what he saw as widespread pork barrelling practices. The reality was, however, that the general exemption had been attached as an amendment to a public-works appropriations bill, and had thus managed to pass in both houses of Congress. Carter could still have stopped the bill with a presidential veto, and under normal circumstances probably would have, but political pressure made him back down. In the midst of trying to pass an education bill and a treaty that would give back the Panama Canal to Panama, he could not afford to make enemies, and even threats made by environmentalist groups that signing the Tellico bill could cost him re-election support was not enough to sway him (Reisner 328-329; Sinclair, "Environmentalists").

After years of fighting to stop the dam, the Authority could proceed with construction unhindered. By mid-November of that year, the last Tellico residents were evicted from their houses, including 75-year-old widow Nellie McCall who had to leave the house that she and her husband had purchased in 1939. Crying as she left the property supported by her daughter, Nellie McCall told reporters that she did not know where she would move to ("Last Tellico"). The

86,000 dollars she was paid for her condemned property seemed a small compensation for losing a life's worth of memories ("TVA's Bitter").

Tellico was in many ways a simulation of earlier TVA projects. On the surface the construction of the dam signaled a return to multi-purpose planning, but the project merely revealed a lack of sensitivity to the needs and objections of the people living in the area. Although parts of the agency's cost-benefit analysis would eventually prove not to be completely erroneous—the area did attract some new industries, about 6,000 new jobs, and spurred the growth of a number of small retirement and golf communities—Tellico was to be the Authority's last major project. When Knoxville journalist Jack Neely visited the area in 2004, many locals still harbored misgivings toward the agency for taking their land and reselling it to wealthy retiree newcomers for several times the original price (Neely). Although it may be difficult for modern visitors to tell Tellico apart from many of the TVA's other projects, as the dam's history shows Tellico lacked the environmental and social vision that had guided the agency's earlier efforts. The return to multi-purpose planning turned out to be a hollow excuse for building a new dam, and the vision of a balanced reintegration of man and nature that had permeated early conservation, recreation, and social planning programs never materialized.

II

Tellico was not alone in exemplifying the breakdown of the pastoralized space's ideological structure. Although the Tennessee-Tombigbee Waterway was built first and foremost for navigation purposes, the project was eventually infused with a sense of pastoral ideology as a response to growing criticism and restrictions imposed by federal environmental policy. In the case of the Tenn-Tom, the notion of integrating nature and culture in a balanced middle landscape became a way of legitimizing a project that was both economically and environmentally problematic, and it was an effort that to some extent succeeded in imbuing the waterway with a new sense of meaning as a recreational landscape.

The idea of connecting the Tennessee and Tombigbee rivers was more than two centuries old. It was first mentioned in the 1760s when the French explorer the Marquis de Montcalm suggested it to Louis XV of France, and was brought up again in 1796 when William Tatham proposed the construction of a canal between the two rivers to the Spanish King. Such a project, Tatham argued, would lessen the friction between the Spanish and the Americans along the Mississippi border. It would allow American traders to bypass the Mississippi River and travel to the sea via the Tennessee and the Tombigbee rivers, removing the Mississippi as a place of contention between the two nations.[4]

Although the Spanish King was not interested in Tatham's idea—in fact he ordered him out of the country out of fear that the Americans might misinterpret his visit—plans for a Tennessee-Tombigbee canal did not end there. In the first decades of the nineteenth century the notion of a waterway resurfaced as part of the attempt to bring steamboats to the upper Tennessee River. The project was investigated by the U.S. Army Corps of Engineers in 1874 and again in 1913, but was deemed too costly. The Corps conducted more studies in 1923 and 1935, but not until 1938 did it decide to recommend the project to Congress. By then, the creation of Pickwick Reservoir in southwest Tennessee had reduced the cost of building the waterway, slightly improving the overall economy of the Tenn-Tom.

However, as with the Tellico Dam the benefit-cost ratio remained problematic. In its review of the 1938 study the Tennessee Valley Authority opted out of commenting on the project's economy altogether, and the Water Resources Committee concluded that although the project might be desirable, its benefits were "speculative" ("Report of National"). Even the Corps admitted to the uncertain nature of the benefits when it concluded that the waterway "could be utilized profitably only for navigation" and that the federal government should only approve of the project if local authorities agreed to bear costs relating to such issues as roads, bridges, railways, and other infrastructural utilities (War Department 142-143).

World War II delayed the waterway's Congressional authorization until 1946 when the Tenn-Tom, along with several other large-scale federal projects, was approved (The Tennessee-Tombigbee Development Authority, "The Tennessee-Tombigbee Waterway").

[4] The Tennessee-Tombigbee Waterway Development Authority, "The Tennessee-Tombigbee Waterway"; Herndon 180.

Rather than a sign of political enthusiasm for the waterway, the Congressional authorization of the Tenn-Tom reflected post-war fears that the domestic economy would return to its pre-war slump now that the war was over. Having thus passed the first of the two Congressional phases of authorization and appropriation, preconstruction planning began. But appropriations for the actual construction were withheld, and when in 1951 the estimated costs of the project had increased markedly from what Congress had authorized, the House Appropriations Committee once more decided that the waterway was economically unjustifiable and cancelled it (Doster and Weaver 184).

The waterway seemed to be a thing of the past, but its proponents persisted. After all, to local supporters the waterway represented industrialization, jobs, and wealth in a region where as many as one third of the population lived below the poverty line. Spearheaded by the enthusiastic Glover Wilkins, the Columbus Chamber of Commerce initiated a local campaign to open the waterway for restudy. A group of southern congressmen joined the effort in 1956, and the following year a restudy of the Tenn-Tom was approved (Stine, *Mixing* 18). In 1958, while the restudy was still being conducted, the governors of Mississippi and Alabama formed an interstate compact. The Tennessee-Tombigbee Waterway Development Authority (TTWDA), as the compact was called, was later joined by Tennessee, Kentucky, and Florida, and had the promotion of the Tenn-Tom as its sole purpose. Funded directly by the tax payers of its member states, with a full-time staff and offices in Columbus, Mississippi, the TTWDA was ready to engage politicians and media on regional and national levels to secure support and promotion for the waterway.

In spite of the restudy showing a benefit-cost ratio of only 1.08-1.0 the Corps, along with the TTWDA, recommended the waterway to Congress on several occasions over the course of the 1960s (Tennessee-Tombigbee Waterway Development Authority, "The Tennessee-Tombigbee Story" 7). In 1966 a new study with a 1.24-1.0 benefit-cost ratio succeeded in securing 500,000 dollars to complete the preconstruction planning, but at the time all public works projects were delayed due to the military escalation in Vietnam and the resulting national deficit ("Waterway Project"). Not until 1970 did the seniority principle in the Congressional committee system combine with President Nixon's Southern Strategy for Reelection to create a political climate that was favorable to the Tenn-Tom project. The waterway's location in the middle of the Deep South could secure the President support in the traditionally Democratic region. Consequently, Nixon included one million dollars in the Corps of Engineers' 1971 budget to begin construction of the canal. Almost 25 years after the project's authorization, the Tennessee-Tombigbee Waterway finally entered the construction phase.

In spite of a broad base of support in Mississippi and Alabama, the two states that stood to gain the most from the Tenn-Tom's completion, on a national level the waterway was a highly controversial subject already before construction was begun. According to critics, its benefit-cost ratio was based on inflated and speculative estimates, the interest rates which the Corps had used to calculate the ratio had been set much too low, there was no guarantee that industry would be attracted by the waterway, and a substantial majority of the Tenn-Tom gains actually benefited coal, grain, and chemical companies rather than the public. Supposedly such corporate savings would be passed on to the consumers in the form of cheaper products and services, but there were no guarantees that they would benefit others than the companies involved (Massey 95-97). Also, the Association of American Railroads (AAR), a longtime competitor of the barge industry, was quick to criticize what it saw as a direct federal subsidy to a rival.

The increased public and political focus on racial inequality also meant that minority organizations and interest groups paid extra attention to the distribution of the Tenn-Tom's social benefits. It was important to ensure that the many poor blacks living along the waterway stood to gain from the new jobs and the rising property values. Consequently, although generally in favor of the waterway, both national organizations such as the NAACP and a host of local groups, including the Minority Peoples Council, monitored and commented on the construction of the waterway and on what was often perceived as discriminatory practices.

Most vocal in its opposition to the project, however, was the environmental movement. The new 1969 National Environmental Policy Act (NEPA) granted the public an unprecedented opportunity to look over the shoulder of federal projects, and due to its size the Tenn-Tom was drawing much attention. When finished, the 234 miles of waterway would flood 40,000 acres and necessitate the removal of 308 million cubic yards of earth and rock, a figure that dwarfed the 220 million cubic yards excavated during the building of the Panama Canal (Massey 96). As Jeffrey Stine notes, the "mere size of the Tenn-Tom riveted the attention of the nation's blossoming environmental movement. As a consequence of moving millions of cubic yards of earth, the project required not only extensive destruction of riverine and wetland ecosystems, but also the joining of two entirely separate watersheds. It was engineering arrogance in a class all its own. Indeed, for many of its opponents, the Tenn-Tom epitomized federal agency insensitivity to environmental values" (Stine, *Mixing* 3).

Early environmental opposition was mostly made up of scattered protests from academics concerned that the Tenn-Tom would destroy the pristine Tombigbee River and its natural habitats. Through letter writing campaigns to the President, the Corps, as well as to members of Congress, biologists, zoologists, and other experts expressed their alarm regarding the potential consequences of the construction (Stine, *Mixing* 91). Among the most vocal was the locally based Committee for Leaving the Environment of America Natural, a group that adopted the memorable acronym CLEAN. Founded in 1969 by members of the Mississippi State University's biology department, CLEAN was originally created to tackle a wide range of environmental issues in the region, but before long the Tenn-Tom came to play a significant role on the group's agenda. Confident that a project of such enormous size would not be able to pass NEPA, CLEAN elected to focus its attention on pressuring the Corps of Engineers to conduct a thorough environmental study before starting the project (Stine, "Environmental" 4-5).

Prior to NEPA and the growing criticism from environmental groups, green concerns had never been a part of the considerations for the waterway. Thus, when Mobile District Engineer Colonel Griffith in 1971 accused environmentalists of "doing an eleventh hour job of superimposing the environment on this project," he not only expressed irritation over the criticism, but also admitted that the environment had not been of major relevance to the Tenn-Tom plans (Stine, "Environmental" 10). Indeed, as late as 1967, plans still existed to include the Tenn-Tom in the Atomic Energy Commission's Project Plowshare. The Plowshare program was initiated in 1957 as an expression of a 1950s technological optimism to promote the use of nuclear detonations for civil engineering purposes: the TTWDA officials believed that it was possible, in theory at least, to excavate the corridor for the northern part of the waterway in a faster and less expensive way by using strategically placed nuclear bombs.

It should be noted that by the late 1950s the environmental problems relating to atomic detonations were well established. In 1954 the Castle Bravo nuclear test in the Marshall Islands had contaminated several inhabited islands, as well as the Japanese fishing boat *Fukuryu Maru* (*Lucky Dragon*) in spite of its position outside of the established restricted zone; in 1955 the United Nations had formed the Scientific Committee on the Effects of Atomic Radiation to study and monitor radiation related problems, and in 1955-1956 five lawsuits were brought against the federal government by ranchers from Iron County in southern Utah, after their sheep had become ill from radiation poisoning resulting from the military's atmospheric tests. The sheep, among many other horrible consequences, displayed burns on their faces and lips from eating radioactive grass. But even if Tenn-Tom advocates had not been paying attention to the political and environmental fall-out of nuclear technology, such problems should have been evident when the Nuclear Test Ban Treaty of 1963 did much to prevent the actualization of Plowshare. Nevertheless politicians from both Mississippi and Alabama protested what they felt to be a federal encroachment on their states' rights, emphasizing at the same time the priority of economic concerns over the environment (Greene 57).

In spite of such formerly cavalier attitudes towards the environment, by 1970 the Corps initiated a pastoral strategy by which it began making minor concessions in order to appease critics. Construction on the waterway, it was promised, would not begin until an Environmental Impact Statement had been filed, and waterway proponents also sought to incorporate environmental and recreational concerns into construction designs as well as into the rhetoric used in describing the Tenn-Tom. In a 1969 presentation to the President and Congress, for example, the TTWDA described the waterway in middle landscape terms as a project that offered the best of both worlds. The Tenn-Tom would "open up a vast area to economic and industrial growth … [as well as] provide a refuge for fish and wildlife and a center for recreational development" (Tennessee-Tombigbee Waterway Development Authority, "The Tennessee-Tombigbee

Story" 5). The presentation further explained that many "waterway projects are objected to by conservationists in our country on the grounds that they upset the world ecology, that is, the total pattern of relationship of all organisms. However, the Tennessee-Tombigbee has never raised objection from any conservationist. On the contrary ... the waterway is expected to provide a new habitat for wildlife; for example, the new flyway for southbound ducks. Esthetically, the Tenn-Tom route, particularly in the area of the impounded lakes, should be an area of natural beauty" (13). It was, of course, an ironic point considering the years of environmental protests and legal battles that lay ahead.

By considering "the total environment and its enhancement for man as well as wildlife," the Corps was hoping to appear ecologically responsible (Stine, *Mixing* 90). By 1971, plans for a part of the waterway, which had originally been designed as a perched canal, were changed to a chain of lakes instead. By leaving the levee on one side of the waterway as planned, to the other side a reservoir was allowed to form according to the contours of the land. This solution, it turned out, was both technically wiser, and aesthetically and recreationally superior, and it even demonstrated a seemingly growing desire to integrate the waterway into the surrounding landscape (Stine, *Mixing* 47-48).

The Corps also altered the design for Columbus Dam in order to appease critics. The original design meant that the reservoir posed a serious threat to a rich fossil bed at Plymouth Bluff only seven miles away. Plymouth Bluff had been the focus of earth science studies for decades, and a research and recreational center had been established there by the Mississippi State College for Women. By late 1971 a new construction site had been located further upstream, and moving the dam the approximately 3,800 feet would save the bluff from being flooded. Since planning was still in the preliminary stage, moving the dam was less troublesome than it would seem, and this way, as Jeffrey Stine writes, Columbus Dam came to serve as a kind of "self-initiated design change made strictly for environmental reasons, which they [the Corps] could hoist up their public relations flagpole as a symbol of their caring and good intentions" (*Mixing* 43-44).

The Corps' newfound environmental sensibilities were mostly symbolic. The promise to file an Environmental Impact Statement before initiating construction, for example, proved somewhat hollow when the Corps' field officers were told to "make every effort to expedite the environmental study" (Stine, *Mixing* 96). To further speed up the process, the study was divided into three phases of which the Environmental Impact Statement only covered the first. Most opponents found this strategy completely unacceptable since it meant that the environmental consequences of the later stages of construction were not even considered. In spite of such criticism, the Council for Environmental Quality accepted the Corps' report. CLEAN's strategy of insisting upon a thorough Environmental Impact Statement had failed to stop the project, and in a change of tactics the group instead joined forces with the Environmental Defense Fund (EDF) with which they had worked before.

The EDF's agenda of strengthening environmental legislation through selected lawsuits was perfectly suited to the Tenn-Tom situation. In 1971 EDF and CLEAN filed a joint lawsuit against the Corps of Engineers, arguing that the waterway would result in the destruction of fish and wildlife habitats and cause irreparable damage to important archeological, historical, and paleontological sites (Stine, "The Tennessee" 9). After winning a preliminary injunction against construction on the waterway, which eventually postponed work on the waterway for more than a year, the plaintiffs were stunned to learn that the Corps' move for a change of venue had been granted. The case had specifically been filed in Washington to avoid the possibility of an Alabama or Mississippi court that might prove partial to the waterway. The move from Washington D.C. had been requested by the TTWDA who wanted the trial held in Mobile, Alabama on the grounds that most of the witnesses came from that area and that moving the Corps' files to Washington D.C. would be a costly affair. Although the new venue was Aberdeen, Mississippi, moving the trial to the heart of Tenn-Tom country was a clear blow to the plaintiffs and an omen of things to come. Not only did the court rule that the Corps did not have to share all of its documents regarding the environmental aspects of the waterway, but it also ruled in favor of the defendants on a motion to limit the issues to be tried, stating that only testimony pertaining to the violation of NEPA would be heard. The economic aspects of the Tenn-Tom, the court concluded, were a matter for Congress to decide (Stine, *Mixing* 118-123).

It was evident from the beginning that the trial was going to be an uphill battle for CLEAN and EDF. Starting in June of 1972, the litigation lasted only eight days. The plaintiffs brought in a host of experts to testify to the inadequacies of the Corps' Environmental Impact Statement: Dr. John Burch from the University of Michigan noted the impact statement's lack of information on the extinction danger which the project posed to as many as 130 snail species and two turtle species; zoology and biology professor from Auburn University Dr. George Folkerts testified that it was possible that competition for food and habitat between the false map turtle from the Tennessee River and the Alabama map turtle from the Tombigbee River might result in the extinction of the latter; a Pennsylvania environmental engineer, Dr. Frank Brown, explained that there was no study of the quality of the water in the new waterway, and that the new reservoirs could potentially reduce, rather than enhance, the area's recreational value; Dr. William B. White from Pennsylvania State University noted the dangers of mixing the two rivers, and criticized the impact statement for failing to mention the changes in flood patterns which the project would cause—changes that were almost impossible to predict but which made flood control benefits a rather speculative asset of the project; and finally, adding insult to injury, professor of sociology at the University of Alabama C. Hobson Bryan explained that the social and economic benefits claimed by the Corps' impact statement were "not documented" and that the "expected economic development may not necessarily occur" (Brumfeld). In spite of the prosecution's impressive use of expert testimony, in August of 1972 the court ruled in favor of the Corps, a ruling which was upheld in 1974 in the Fifth Circuit Court in New Orleans (Stine, *Mixing* 125-126).

Opponents of the waterway did not give up. The Association of American Railroads had monitored and opposed the waterway since 1934 when the association was formed, and had even provided witnesses to testify against the canal at Congressional hearings (Stine, *Mixing* 15). The Louisville & Nashville Railroad (L&N), a member of the AAR, had offered consultant and financial aid to CLEAN during the early part of the fight against the Tenn-Tom, but had declined an offer to join the earlier litigation, preferring instead to work in the background via lobbying efforts and criticisms of the project's economy. But after the Corps had won the first trial, it was becoming clear to the L&N that a more aggressive strategy was needed. Thus, in

December of 1974 the railroad offered to join in a new lawsuit (Stine, *Mixing* 124-143).

With the L&N on their side the EDF and CLEAN had strengthened their case considerably. The railroad added not only substantial political muscle but also weighty economic arguments to the environmental issues. In 1976 the three partners filed a parallel lawsuit. As opposed to the first trial, the new litigation centered on three distinct issues: the legality of the Corps' design changes, which had altered the project from the construction for which Congress had appropriated money; the legality of the benefit-cost analysis; and whether or not Tenn-Tom was in compliance with NEPA regulations.

The matter of the Corps' authority in redesigning the waterway was settled first. In 1979 the court ruled that "the day for battle on the authorization issue had come and gone," concluding that the plaintiffs had unduly delayed in bringing the case to trial ("Appeals Court"). The verdict was upheld in the federal appeals court, and not long thereafter the district court also dismissed the benefit-cost issue and the NEPA violations as being outdated. The appeals court changed those verdicts and ordered the Corps not only to update its Environmental Impact Statement in accordance with the new designs but also to begin using a higher interest rate in its benefit-cost analysis. Issuing a temporary injunction against construction, the appeals court might have put up an effective road block for the canal, but the court subsequently allowed for the district court to delay that injunction in specific instances, which effectively created a legal loophole that allowed work on the Tenn-Tom to continue (Stine, *Mixing* 212-215).

In spite of having once again survived litigation, the Corps' situation was far from ideal. By the mid-1970s it came under attack from minority groups for not living up to its promises of racial equality, and the price tag on the waterway kept increasing, from 346 million dollars in 1971 to a final cost of two billion dollars. From 1975 to 1976 alone, the estimated cost increased from 815 million to 1.36 billion dollars, lowering the benefit-cost ratio to 1.1-1.0 (Stine, *Mixing* 156). However, according to independent economist and long-time critic of the waterway Paul E. Roberts, the actual benefit-cost ratio was no more than 0.75-1.0, an estimate which meant that tax-payers lost 25 cents for every dollar spent on the project (Gibson).

Furthermore, the second trial revealed documents proving that the Corps had deliberately misled Congress about the actual costs of the project for fear that appropriations would be withheld. Instead of reporting a 500 million dollar cost miscalculation to Congress, the Corps hired the A.T. Kearney Management Consultants, Inc. to restudy the traffic projections in order to boost the possible benefits. But to critics the Kearney report's estimated total of 28 million tons of cargo transported on the Tenn-Tom during its first year of operations seemed almost impossibly high. Of the estimated 28 million tons, 7.2 tons were metallurgical coal from northwestern Georgia, northeastern Alabama and southern Tennessee, yet at the time those areas were producing only 1.5 million tons for export a year (Massey 97). The proponents' argument that due to the energy crisis much more coal would have to be produced, and transported, in the future did not add up. According to transportation economist Joseph L. Carroll, who had originally been hired by the Corps to analyze the waterway project in search of overlooked benefits, the agency "had coal coming from parts of the country that didn't have that much coal in the ground" (Peterson). Also, one of the supposedly major canal users in the Kearney report had gone out of business, and others never had intentions of using the Tenn-Tom (Sinclair, "Spurious"). The trial also revealed that the agency had buried two unauthorized canal locks at a cost of 107 million dollars in their budget, and that benefits had been raised by 60 million dollars a year through fictitious calculations in order to boost the benefit-cost ratio ("The Engineering of Deceit"; Kovacs and Guge).

Such criticism made it more important than ever for the Tenn-Tom supporters to portray the project in a positive light. In 1975 plans for four recreational areas with picnic facilities and boat ramps were launched, and the following year plans for an environment and wildlife studies center were unveiled to the public (Fields; Lynn). In 1977 the Corps made public their intent to create a 50,000 acre recreational area that would stretch from Amory to the Tennessee state line ("Huge Recreation"). It was an effort that would result in what one paper called "an outdoor paradise of lakes, parks, game management and wildlife enhancement areas" that would include "natural, aesthetic, archeological and historical resources along the waterway" ("Tenn-Tom").

Not everyone accepted the Corps' new green plans at face value, however. In a 1976 article in *Smithsonian* magazine, Don Moser, former journalist for *Life* magazine, wondered whether the agency's new attitude was more than skin deep. Asking if "the Corps [is] undergoing a transformation, changing from an earth-moving agency to an environmental watchdog" (44). Moser skeptically answered his own question by concluding that the Corps' environmental agenda seemed superficial, a change of procedures without a "corresponding change in values" (50).

The Corps' attempts to act on environmental criticisms on occasion also brought it into conflict with local residents, but such conflicts, more often than not, hurt the environmentalists rather than the Corps. In 1981, for example, the Corps responded to a recommendation by the U.S. Fish and Wildlife Service to recreate some of the natural habitats lost in the construction of the waterway by setting aside 97,000 acres for wildlife preservation. The plans caused uproar among local residents who felt that they had given up enough land to the Corps. Three public hearings about the plans were held in Livingston, Alabama, as well as in Fulton and Columbus, Mississippi, and all three meetings resulted in jeers and angry words from the attending crowd (Bailey; "Landowners"). At the meeting in Columbus, Chester McConnell, a visiting representative from a private environmental interest agency in Tennessee, was verbally abused and threatened by the crowd, while all attempts at communicating the environmental concerns to the audience were virtually ignored. Even comparisons with the dire environmental situation in the southern part of Mississippi—an area which, according to a representative from the Fish and Wildlife Service, had been in much the same situation twenty years earlier and had elected to do nothing—meant little to the landowners, and the project ultimately was cancelled (Mould). That result allowed the Corps to document that it had attempted to comply with environmental recommendations, while at the same time supplying the agency with arguments against the arguments of the environmentalists.

The Corps' second legal victory meant that, by 1983, all that was standing in the way of completing the waterway was an increasingly hostile Congress. On a few occasions the project only barely survived a close Congressional vote on the annual appropriations, and only arguments that the waterway was so close to completion that it would be a waste of tax-payers' dollars not to finish it, secured the needed majority of votes (King). Finally, in January 1985, the Tennessee-Tombigbee opened for traffic, ending 40 years of political struggle. But the waterway did not open to the kind of success that many proponents had been dreaming of. Economically, the Tenn-Tom fell far short of its estimated benefits, and its first years of operations were reviewed by the national media with scathing criticism. As Cass Peterson of *The Washington Post* wrote after visiting Amory, Mississippi, in late 1986: "The Tennessee-Tombigbee Waterway's grand opening in 1985 is commemorated in framed photographs on Mayor Tom Griffith's office wall here: beaming congressmen, a smiling beauty queen, a spray of fireworks reflected in the dark waters of the canal that was to be Amory's pathway to prosperity. ... Nearly two years later the fireworks have long since gone out and the prosperity has yet to arrive. Amory's industrial park sits vacant. The town's port is incomplete and idle, partly because of construction difficulties and partly because there appears to be no pressing need to complete it" (Peterson).

For the Tenn-Tom's supporters the reality was indeed depressing. By the end of the first year of operations only 1.7 million tons of cargo were transported via the waterway, a mere 6% of the tonnage that had been predicted in 1975, and only about 11% of the careful estimate made during the summer of 1985 (Schmidt). In 1990 only 4.7 million tons out of a projected 32 million tons were transported via the waterway, and the number of people in Pickens County living below the poverty line had increased from 27.1 to 28.9 percent in the ten-year period from 1979 to 1989 ("Atlanta"). Two years after the grand opening with thousands of visitors flocking to the waterway, the Tenn-Tom's anniversary went "unheralded," and in 1989 the *Clarion-Ledger* called the Tenn-Tom a "$2 billion question" and noted that the "long ballyhooed economic windfall used to justify the work has for the most part not materialized" (Page; Huffman and Rejebian). Over time, other estimates also turned out to be erroneous. Predictions that the population of Tishomingo County in northern Mississippi would grow from 14,940 (1970) to at least 60,000 (1990) was overly optimistic (Weiler and Downing). In fact, according to the U.S. Census Bureau, by the year 2000 only 19,163 lived there.

Continuing the pastoral strategy even after the waterway's completion, proponents of the Tenn-Tom persisted in portraying the construction in terms of environmental enhancement. The TTWDA brochure *Touring the Tenn-Tom Waterway* describes it as "an outdoor enthusiast's paradise" and further emphasizes that the Corps, by Congressional authorization, was managing approximately 92,600 acres of project lands, and might purchase an additional 88,000 strategically located acres, for "wildlife mitigation activities" including "habitat management for game, non-game and endangered species." Likewise, the story of the Tenn-Tom is given a positive environmental spin in the pamphlet *The Tennessee-Tombigbee Waterway: Twenty Years of Economic Progress—And Still Counting.* Under the headline "Environmental Quality" the brochure explains how the Tenn-Tom "was the first large public works project constructed under the National Environmental Policy Act" and that the waterway "demonstrates how a complex project can be built in an environmentally compatible manner."

Fig. 12: The Stennis lock and dam on the Tenn-Tom Waterway near Columbus, Mississippi. Photo by author (2007).

The Corps also sought to present the waterway as less intrusive through the appearance of the new recreational areas. Striving for a natural look, many of the structures were made of wood, adding a rustic feel to the area: wooden handrails frame parts of the nature trails, footbridges allow visitors to cross smaller streams when hiking, and pavilions afford visitors a chance to relax in scenic surroundings. Introducing such features at the campsites has created a sense of anti-modernity that has softened the image of the new man-made canal and made it appear more authentic.

Further blurring the distinction between the authentic nature experience and the constructed environment of the Tenn-Tom, beginning in the 1970s proponents of the waterway consistently began portraying it as both an original and natural landscape. A set of promotional materials by the TTWDA collected under the title "Tennessee-Tombigbee Waterway—Your Waterway" refers to the Tenn-Tom region's "almost virgin countryside," and a promotional brochure for the De Wayne Hayes Recreation Area talks about "the natural beauty of the Tennessee-Tombigbee Waterway."[5] The Town Creek Campground is likewise described in a tourist brochure as complete with modern facilities "while its setting on the banks of the old Tombigbee River is nostalgic of earlier days" (US Army Corps of Engineers, "Town Creek"). In several recreational areas, campgrounds without water or electricity were established for those visitors "who really want to get back to nature" (US Army Corps of Engineers, "Piney").

Ironically, the recreational use of the Tenn-Tom exceeded all expectations, even surpassing navigation as an economic factor. Half a year after the waterway opened for use, it was clear that anglers, hunters, and pleasure boaters by far outnumbered the commercial uses of the waterway, and by the early 1990s an estimated six million visitors had frequented the Tenn-Tom lakes ("Recreation"; "Atlanta"). In many places along the waterway, marinas, restaurants, hotels, and bait shops prosper where the industrial infrastructure stands unused and unfinished. In all fairness, to many communities recreation became their best chance to harvest any Tenn-Tom related economic benefits, and focus has shifted to promoting the waterway as a

---

[5] The Tennessee-Tombigbee Waterway Development Authority, "Tennessee-Tombigbee Waterway – Your Waterway"; US Army Corps of Engineers, "DeWayne".

recreational landscape (Peterson). Thus, in spite of the historic reality that environmental issues have been superimposed on the project by a reluctant agency, the Tenn-Tom was given new meaning and a second life as a recreational attraction.

The Tenn-Tom remains a contested project due to its economic failure and the environmental destruction that followed in its wake, and although the man-made waterway has become popular among recreationists, the recreational and environmental benefits that were associated with the Tenn-Tom emphasize the breakdown of the pastoral ideology rather than its continuation. The introduction of pastoral designs through environmental and recreational initiatives was primarily due to the continuing fiscal and environmental troubles that threatened to stop the project for good. As the following chapter will show, projects such as the Tellico Dam and the Tennessee-Tombigbee Waterway have helped free the pastoral from its former context and significance, paving the way for a hyper-pastoral space where the distinction between the real and the represented has disappeared.

Five

"You don't get that authenticity at Disney!"[6]: Dollywood, Jack Daniel's, and the Emergence of the Hyper-Pastoral

I

Early in his novel *Flood* Robert Penn Warren describes a small engineered gardenscape outside of the newly built Seven Dwarfs Motel. Seemingly an idyllic and peaceful place constructed for the pleasure of the visiting tourist, to protagonist Bradwell Tolliver the blend of nature with man-made constructions to create an absurdist fairy tale environment is profoundly disturbing:

> The big sycamore by the creek was gone. The willow tangle was gone. The little enclave of untrodden blue-grass was gone. The clump of dogwood on the little rise across the creek—now that, too, was gone.
>
> But the trouble was not so much what was not there. It was what was there. The creek was there, but it flowed decorously between two banks where stones were mortised into the earth; and on a boulder a cement frog, the size of a young calf and the color of Paris green, with a mouth gaping as richly bright as a split liver on a butcher's block, crouched. On another boulder a gnome, dwarf, brownie or some such improbability, with a cement beard painted snow-white, sat studiously fishing. The line with which the creature was fishing was a real line. It wavered with the motion of the stream. The water in the stream looked real. But the water lilies were definitely cement. (Warren 4)

---

[6] Stokes, "Welcome to Dollywood".

The simulated frog and water lilies undercut the sense of authenticity that Tolliver associates with the rural South he remembers from his childhood. Instead of the rustic scenes he expected, he finds signifiers of rurality designed to appeal to modern consumers hungry for nostalgia. The comparison of the cement frog to a young calf fittingly hints at the biblical tale of the molten idol created and worshipped by the impatient followers of Moses. Part of a commercialized space, the frog symbolizes the new consumerism that, by the 1960s, had become the primary object of America's secular worship.

Tolliver has returned home to make a movie about the idyllic Fiddlersburg as he remembers it and has described it in a novel, but he soon finds that the new dam has transformed the area physically, economically, and culturally. The Seven Dwarfs Motel represents a society in which everything has been reduced to customer appeal, and where even the past has become just another marketable product. The motel owners cater to a public hungry for the nostalgic appeal of the pastoral, and even the fake southern accent and the stereotypical subservient attitude of the young black man working at the motel is a matter of simulation. Similarly, a large neon billboard at the side of the road features a black minstrel-type face and the text "Breakfast Served in Cottage. Tennessee Smoked Ham and Red Gravy. Yassuh, Boss!". Selling a sense of history through its references to cottages and old-timey recipes, the billboard also mockingly reinterprets a hurtful part of the region's racist past in order to sell an image of anti-modernity and Southern charm (Warren 4).

Published in the early 1960s, *Flood* is interesting in part due to its observations on the transitional nature of the post-TVA South, but also because Warren hints at the emergence of a hyper-pastoral space dominated by empty representations. As the motel's name and the inclusion of the over-sized frog and the gnome suggest, the history represented there refers to a fiction. In much the same way, Tolliver himself engages in a process of re-fictionalization as he seeks to convert his own literary portrayal of Fiddlersburg to film.

With the breakdown of the ideologically-driven pastoralized landscape, the pastoral has become increasingly decontextualized and removed from its historic frame of reference. In places such as the Chattanooga Choo Choo, Dollywood, and the Jack Daniel's distillery in Lynchburg the pastoral has become an empty and fragmented experience, not unlike the copies of the Manhattan skyline or the

Eiffel Tower that can be experienced in Las Vegas. As opposed to the pastoralized landscapes of the Biltmore Estate, the TVA's early programs, or the Museum of Appalachia, which constitute more or less romanticized attempts at copying the historic pastoral into a functional contemporary reality, the hyper-pastoral is dictated by its own internal logic. In California the small town of Solvang, for example, is structured around a conspicuous and highly stylized sense of Danish ethnicity that includes windmills and fake half-timbered homes, not a few of which have plastic storks perched on the roofs (Jones, "The Danish Soul"). Today the almost exclusive aim of the town is to attract tourists, and with around 1.7 million visitors every year its appeal is undeniable. In 2007 the pastoral narrative cultivated at Solvang was even caricatured on an episode of *The Simpsons*.[7] While picking up his friend Milhouse's Danish uncle at Springfield airport, Bart Simpson witnesses the arrival of a plane from Solvang Air which turns out to be a half-timbered home with wings.

Solvang was originally created in 1911 by a small group of Danish settlers from the Midwest. After World War II, and following a celebratory article in the *Saturday Evening Post* in 1947 which heralded the charms of the small town, the town began to develop into a tourist attraction. In the 1950s, as tourists began visiting the small California town, a process of "Danish-provincializing" commenced on an ad hoc basis. "Danish-provincializing" referred to the "Danish provincial" style that was introduced by a local man, Ferdinand Sorensen. Sorensen, who had just been to Denmark for the first time in his life, felt so enamored by what he experienced as the pastoral feel of the small Scandinavian country that he decided to build his own home in a similar style (Linde-Laursen 788-790). By the 1960s the marketing of the town had become such an important economic issue that a "Danish provincial" design strategy was adopted, and by the 1970s and 1980s older structures were being demolished "for the sake of Solvang's modernization and danefication" (Linde-Laursen 790). It is interesting that whereas the builders of the original structures, such as the Solvang Youth School from 1911, did not rely on public signifiers of their Danish heritage, newer buildings increasingly relied on empty but highly visible identity markers, such

[7] The episode "Little Orphan Millie" (season 19, episode 6) was aired on November 11, 2007.

as architecture and flags, to signal a Danish heritage; an ironic development considering that while the danefication of Solvang increased, the percentage of residents claiming Danish ancestry decreased. In fact, less than twenty percent of the current Solvang population can claim a Danish heritage, and it is a fair guess that not many of the inhabitants have actually visited Denmark (Linde-Laursen 796). Rather, behind the seemingly innocent pastoral facade of Solvang, many of the town's businesses run on Mexican labor, a fact which hardly fits the dominant narrative of Danish pastoralism. Furthermore, the imposition of an ethnic Danishness upon an area historically peopled by Chumash Indians and people of Hispanic or Chicano descent is problematic (Linde-Laursen 782).

Interestingly, and perhaps precisely because the decontextualized hyper-pastoral has no historical point of reference, displays frequently involve antique technologies like steam driven trains. In the hyper-pastoral framework the technologies that landscape advocates once considered the very antitheses of nature have become vehicles of reclamation that function to bring the tourist back into an idealized past where the machine no longer exists in opposition to the garden. The Tennessee Valley Railroad Museum (TVRM) in Chattanooga is an example of this tendency. The TVRM was originally formed in 1959 by a group of railroading enthusiasts who were concerned about the disappearance of steam locomotives and passenger trains from the American railroads. The group was formally recognized as a non-profit, educational organization in May 1961 after public responses to two steam train excursions—the "Mountain Goat Special" and the "Oz Special"—had convinced the organizers that there was sufficient interest in the preservation of antique steam trains. The group officially organized as a local chapter of the National Railway Historical Society, and already in April 1961 members were negotiating with the city and the railroads over a permanent site and about what to display there ("Steam Train"). By January of 1962 the groundbreaking ceremonies for the new Warner Park Museum were held. The site had been donated by the city, and the planned excursion to Copper Basin was planned in cooperation with the Southern Railroad and the L&N ("Railroad Unit").

Although the museum's focus is on the technology of the steam train, over the years its activities and promotional materials have served to inscribe that technology into a romantic narrative where railroad and landscape merge. Reminding visitors of the supposedly romantic days of steam railroading, the TVRM refers to a largely imagined past that visitors can relive through antique train rides into the scenic landscape. In the promotion of the 1962 "Copper Basin" Special tour, for example, the text reads: "Wild flowers and mountain laurel will be in bloom along the beautiful Hiwassee gorge, accessible only by train," and continues by explaining that the "route we will take traverses some of the most beautiful and spectacular mountain scenery in the South, varying from lush farmland valleys and plains to rugged mountain gorges and roaring streams" (Tennessee Valley Railroad Museum). The idea that the part primeval and part pastoral landscapes of the Hiwassee gorge should be accessible only by train suggests that taking "The Copper Basin Special" serves as the one way for the tourist to reenter a past that has otherwise been lost, a narrative which effectively posits the train as part of the cultural simplicity for which many tourists long.

The power of the train as a signifier of anti-modernity is evident when one considers two similar events involving one of the museum's old locomotives. On June 6, 1964, No. 4501 traveled towards Chattanooga to take its place at the new railroad museum. A classic locomotive from the Southern Railway System, the 4501 hauled three passenger cars full of excited travelers. Along the way the train was greeted by thousands of people coming out to wave and take pictures. It was saluted by passing modern diesel trains, and during parts of the journey where the highways paralleled the train tracks, it was even accompanied by lines of curious motorists (Scheider). Clearly the appearance of the old locomotive engaged the public imagination.

For some, however, waving and taking photographs of the old train was not enough, and at Spring City the local chamber of commerce, upon learning that there was no stop planned in their town, dressed up as bandits, staged a hold-up, and passed out lunches to the passengers ("The Holdup"). The incident was repeated two years later when the 4501 was held up once more on a 14 hour journey to southern Indiana as a part of that State's Sesquicentennial celebrations. Having been thoroughly renovated, the train—once again fully loaded with passengers—was stopped to the sound of Indian war-whoops, as a group of participants in an outdoor drama boarded the train armed with tomahawks, passing out promotional material on their local "Daniel Boone" play ("Odyssey of"). The convergence of staged histories in the two hold ups of the 4501 suggests the strength and appeal of public nostalgia associated with the antique locomotive. In both cases the evocation of regional pasts by the appearance of the old train seemingly provoked what might best be described as a desire not merely to observe but to people that historic landscape. Assuming outlaw and Native American identities for dramatic effect in advertising-related gimmicks, the hold-ups complemented the train ride and brought into focus the plasticity of the pasts that were being performed by those holding up the train and the train enthusiasts alike.

A similar idealization of antique technologies can be experienced in downtown Chattanooga, where the Southern Railway's famed Terminal Station has been preserved in the form of the Chattanooga Choo Choo hotel and restaurant complex. Complete with flower gardens and antique trains refashioned into hotel rooms, the new complex not only constitutes an effort to renovate the Terminal Station itself, but it also infuses the history of the railroad with a pastoral theme, displaying "history" as just one among a host of family-oriented attractions. Adding the famous Glen Miller song as a nostalgic hook, the Chattanooga Choo Choo is—as the hotel website proudly exclaims—a train, a song, and a hotel.

Terminal Station was built by the Southern Railway Company in the early twentieth century. The Southern Railway Company was formed on July 1, 1894 as a consolidation of the two regional railroads, the Richmond and Danville Railroad System and the East Tennessee, Virginia & Georgia Railway Company. Throughout most of the 1880s and the early 1890s the East Tennessee, Virginia & Georgia Railroad Company was engaged in litigation with the State of Georgia and a number of other railroad companies over the decision to expand Chattanooga's existing Union Station. Unhappy with the plans and the amount of money that an expansion would ultimately cost, the East Tennessee, Virginia & Georgia Railroad Company instead attempted to force a sale of the station by bringing the case to the courts. Apart from a long and costly trial, however, the attempt proved fruitless. By the mid-1890s—the plaintiff's role had been taken over by the Southern Railway Company in the meantime—the Tennessee Supreme Court ruled in favor of the state of Georgia, declaring that the Southern Railway Company's status was that of a lessee rather than an owner, and therefore it had no vested rights in the actual building (Govan 377-378). In light of the state Supreme Court's ruling and the growing market that followed the Southern Railway's constant expansions throughout the 1890s, it was only logical that Southern Railway would eventually look to construct a station of its own. And since Chattanooga was one of the most important towns in the Southern railroad system, in 1904 the company purchased the Stanton House, a once renowned hotel which had fallen into disrepair, and demolished the five-story, 100-room building in order to pave the way for a new station.

In choosing a station design, the Southern Railway Company received proposals from both established and promising young architects, but one design in particular called attention to itself. It was submitted by a young New Yorker, Don Barber, and embodied the splendor of the Victorian era with its rich ornamentation and classic style. Barber's designs had won him a prize at the famed Beaux Arts Institute in Paris only four years earlier, and on meeting with the president of Southern Railway, he was soon informed that the job was his. The exterior of his proposed station was perfect, he was told, and the only alteration to the design was that it would be preferable if the interior of the station echoed that of the famous National Park Bank in New York City, a request to which Barber happily agreed ("Terminal Station").

A lot of attention went to the design of the building as well as to the exterior grounds in order to appeal to the public both in terms of aesthetics and user-friendliness. The station was built on one level unlike the stations in, for example, Birmingham and Atlanta, a design feature which saved travelers from having to climb the stairs while carrying heavy suitcases. The baggage, mail, and express departments were placed adjacent to the main building, and only the newest technology was used. Thus, all of the switches for the main yard were electrified and operated from one central switch tower. The area in front of the new station was graded and covered with artistic flower beds and, adding a final touch of elegance in completing the impressive construction, a landscape gardener was hired to ensure that the overall look of the new facade lived up to expectations ("New Station".

Ever since the plans for the grand construction were revealed, there had been a general sense of anticipation with regard to the new station. Local papers had followed its 18-month construction closely, and on opening day, December 1, 1909, a crowd of 500 excited bystanders turned out for the ceremony (Shearer). The impressive red brick arch, which was said to be the largest in the world at the time, awed the spectators, and the impressive interiors of the main hall were among the first impressions most travelers got of Chattanooga. The building was quickly proclaimed the finest in the city, and such was the excitement that a traveling salesman from Knoxville, R.N. Bearden, made sure to purchase the first ticket, just, as he explained, to be able to say that he was indeed the very first customer at the new depot ("Terminal Station").

For decades Terminal Station served its purpose both as a train station and as a tourist attraction. The station was even included in Glenn Miller's popular 1941 song "Chattanooga Choo Choo" from the movie *Sun Valley Serenade*, and although the phrase was originally coined by a reporter to describe the early small wood burning locomotives that traveled South from Cincinnati in Ohio, the popularity of the song in the U.S. and among the troops fighting in Europe soon made the Chattanooga Choo Choo and "track 29" household names in most of the Western world.

But whereas a large portion of the domestic transport during the war had gone by train, in post-war America the railroads came under increasing pressure from the automobile and air travel industries. It was a losing battle. By August 1970 the last passenger line, the "Birmingham Special," was removed from service, and boarding up the windows of the station, the exquisite interiors, once the pride of the city, were left to collect dust. As one paper commented, the famed Terminal Station appeared to be "doomed to eventual demolition" ("Passing from").

Two years later, a group of local businessmen recognized the potential value of the old building, and Terminal Station was given a second life. On February 12, 1972, *The Chattanooga News Free Press* first reported the plans to turn the old train station into a 24-acre "multi-million dollar restaurant-motel complex" built around a "Gay '90s railroad motif" ("Restaurant at"). The new complex would house a Hilton Motor Inn and a working model railroad, turn-of-the-century tourist shops, a railroad museum, renovated train cars refashioned into hotel rooms, and street cars to transport visitors around the grounds on guided tours. Behind the plans to build the new complex was the Chattanooga Choo Choo Co. Inc., a consortium of local businessmen investing a total of four million dollars in the renovation of the station.

The plans were highly praised from the beginning by city and county officials. At the groundbreaking ceremonies on April 29, Mayor Robert Kirk Walker called the project one "of the greatest physical developments in Chattanooga history," and according to the president of the Greater Chattanooga Chamber of Commerce, Edward M. Copper Jr., the Chattanooga Choo Choo would "make tourism secure as a basic `industry′ in the community" (Station Project"). Furthermore, the group's declared goal of paying specific attention to historic detail in restoring the station was praised as one of community pride, showing above all the "desire to preserve all of Chattanooga's rich railroading tradition" ("Station Project").

In spite of the preservationist goals behind the project, the Chattanooga Choo Choo was a business, and history was one of the attractions being sold. Consequently, it was a promotional boost for the hotel complex when, in early 1973, Terminal Station was included in the National Register of Historic Places (Baker, "Tennessee"). Yet history as a theme was only interesting as long as people came to experience it, all the while spending their tourist dollars at the complex. To attract visitors a new 50,000 dollar sign was placed on top of the building to be "seen for miles" around. Depicting a train with the words "Choo Choo" on it, the sign included "neon lights and hundreds of small light bulbs that will blink to give the illusion of the locomotive moving and smoke billowing" ("Getting Up"). The emphasis on the tourist experience also meant that the narrative that was being sold at the complex had to be one that appealed to a wide range of people. The hotel itself was built up around a "gay '90s" theme, and the antique train cars, now refashioned into hotel rooms, were placed on two parallel tracks amidst impressive flower beds, suggesting the kind of romantic past that would resonate with the nostalgia of the late-twentieth century tourist.

The garden area, with its fountains, statues, walking paths, small tropical trees, and a host of different flowers, including 4000 red and white begonias and 6000 marigolds, create an appealing atmosphere for those living in the train car rooms (Baker, "Flowers"). But for the many thousands of visitors not staying at the hotel, the garden also plays a vital part in evoking images of an imaginary, peaceful, and environmentally balanced past. Far from the Leviathan that Horace Kephart witnessed despoiling the Tennessee forests and consuming the wilderness, rather the train is displayed as an integrated part of the gardenscape (see cover of this book). Another way in which the Chattanooga Choo Choo Co. Inc. has sought to cultivate the image of the past is through the so-called Septemberfest. Starting in 1977, the company began inviting craftsmen from all over America to come during the month of September to demonstrate, exhibit, and sell their work from booths set up in the flower garden (Mitchell). Along with the renovated station, the antique train, and the blossoming garden, the handicrafts displays can signify a return to what one newspaper referred to as "the good old days" ("Crafts Fair").

Terminal Station makes a wonderful example of what *Southern Living* editor Philip Morris has called "adaptive-use preservation"; a type of preservation in which historic sites are preserved and used for other purposes than what they were originally intended for (Morris). Like the Tennessee Valley Railroad Museum, the Chattanooga Choo Choo restaurant-hotel complex exemplifies a narrative in which antique technologies and a blossoming gardenscape merge into a vision of middle landscape bliss. It is a nostalgically-driven approach to a largely imagined past, and like most hyper-pastoral places it has little referentiality beyond the images it uses.

II

Dollywood is one of the Tennessee Valley's largest and best known tourist attractions. Since 1986 the amusement park, partly owned by singer Dolly Parton, has been in the business of peddling its version of regional history, and with more than two million visitors a year the park's appeal is unquestionable. As a Danish journalist recently noted, the park reminded him of a mix between Tivoli and the open-air museum the Old Town in Aarhus (Pedersen). Other reviewers have likewise found Dollywood a less stressful setting than the average theme park, insisting, for example, that it is neither the rides nor Dolly herself that are the central aspects of Dollywood; rather, it is the theme of life in the Appalachian Mountains that matters (Smyth, "Hello"). As Alison Stokes of the South Wales Echo wrote after visiting the park in 2008:

> While you might expect a theme park dedicated to the country's most flamboyant star to be as plastic and as

> kitsch as the lady herself, you'd be wrong. Nestling in the shadow of the Great Smoky Mountains National Park, it's a 130-acre theme park for country folk, dedicated to preserving those simple ways. Sure it's great fun with shops, theatres, fairground and water rides and some of the best white-knuckle roller-coasters anywhere. But it's also a thriving heritage museum where you can watch craftsmen carving fiddles and horse carriages, tanning leather belts and making candles and soap. There's even an old school house and a chapel on site, an American eagle sanctuary and a real "Casey Jones" steam train. You don't get that authenticity at Disney! (Stokes)

Certainly that is exactly the type of response that the owners of Dollywood are trying to produce. As Parton stated in her opening day speech, her goal for the park was "to preserve the Smoky Mountain heritage so people can come here and see what we are really all about, rather than have some Hollywood bunch portraying Tennessee mountain people" (Smyth, "Hello").

But is Dollywood that different from most parks? Disneyland's Frontierland, for example, also includes historicized exhibits, and Dollywood is, in fact, only the latest development in a long line of Pigeon Forge amusement parks that have all sought to sell stylized images of Southern and Appalachian history and culture. The first to open its doors on the grounds that would eventually house Dollywood was the Rebel Railroad theme park. From 1961 through the better part of the decade, the park featured "a coal-fired steam train, a general store, and a saloon," a minor display of regional craftsmanship in the form of a working blacksmith, and the park's main attraction: a five-mile train ride during which visitors had to defend themselves against an attack by Union troops (Uhlenbrock). Leaving aside the curious location of a rebel theme park in a county that in 1861 voted "no" to secession with 1302 votes against and only one in favor, the staging of Union troops as the enemy more than a century after the end of the Civil War and at a time when the country was in the grips of a raging civil rights struggle, must indeed have seemed a truly bizarre experience to many visitors.

In 1970 the park was sold to Art Modell, owner of the Cleveland Browns football team, who styled his Goldrush Junction theme park after the Old West and included attractions such as a wood shop, a grist mill, and a sawmill. As Modell soon realized, however, people did not travel to the Great Smoky Mountains to experience a copy of the Wild West, and shifting the park's focus to a Smoky Mountains theme, Modell instead sought to replicate many of the stereotypes found in popular TV shows like "The Beverly Hillbillies", "The Andy Griffith Show," and "Hee Haw" (Martin 214-215). By using stereotypes and a milieu that visitors could readily recognize from their TV screens, Modell hoped to appeal to the cultural nostalgia that made CBS's rural shows such successes throughout the 1960s. But financially Modell's park was not a success, and in 1976 he sold it to the Herschend brothers, who had made a fortune on the tourism trade in Branson, Missouri. The Herschends owned the Silver Dollar City theme park, an amusement park which also relied heavily on country and hillbilly metaphors; in fact, several episodes of "The Beverly Hillbillies" had been filmed at the Silver Dollar City in Branson (Smyth, "Hillbilly"). The new park, Silver Dollar City Tennessee, also based its attractions on a mix of hillbilly stereotypes and regional craftsmanship and it actually featured many of the same attractions that could be found in Branson.

Although the Silver Dollar City Tennessee was a moderate success, when in 1985 the Herschend brothers heard about singer Dolly Parton's interest in getting into the growing tourist industry of her home county, they contacted her with an offer to make the singer a minority partner. The park was renamed Dollywood and expanded for six million dollars. With Dolly Parton attached, the park's owners focused more than ever on furthering the country theme. Craftsmanship was an important part of the effort to emphasize rurality and local traditions. In the Craftsman's Valley part of the park, visitors could watch expert blacksmiths, potters, and carriage makers at work. Over the years the park has even run workshops to teach interested participants different skills such as rug crocheting or the making of "Victorian paper dolls, wearable art, pottery, dulcimer-making, stained-glass art and watercolor painting" (McDonald, "Dollywood").

Craftsmanship has long played a significant role in the marketing of Southern culture. By the late nineteenth century regional crafts had found their way into Southern resort areas, driven primarily by well-meaning visitors who imagined the Mountain South as a temporal vacuum undisturbed by modernity and who saw handicrafts as a preservation-worthy expression of this anti-modernity (Martin 75-76). In fact, the birth of the Southern arts and crafts movement is often attributed to Frances Goodrich, a Yale graduate who moved to Buncombe County in North Carolina in 1890. Running her Allanstand Craft Shop out of Asheville, Goodrich became central to the establishment of the Southern Highland Craft Guild in 1928 (Southern Highland).

Yet whereas the establishment of the Southern Highland Craft Guild was a way of honoring a trade and at the same time allowing the women involved a means of economic and social independence, Dollywood displays of traditional handicrafts exist merely as part of the larger effort to sell visitors a romantic and static image of mountain life. The addition of an inspirational life-size wooden display of a carpenter and a young boy with the inscription "If you work with your hands, your mind and your heart .... you are a craftsman," certainly implies that handicrafts at Dollywood are less about a mountain reality of few luxuries, and more about a late twentieth century nostalgia for cultural simplicity and a lost romanticized era endowed with a sense of greater coherence and purpose. Although the craftsmen at Dollywood are undoubtedly skilled, the rows of home-made products found in the nearby tourist store ultimately serve as reminders of the economic realities that govern the park's rustic displays.

The Herschend brothers were counting on the affiliation with the local country music star to boost the park's attendance figures, but the reality of the park's opening season exceeded all expectations. Dollywood experienced a 100% increase over the previous season's 750,000 visitors, causing the park's director of marketing to speak of "the Dollywood Phenomenon" ("Dollywood"). Dollywood's early success meant that an immediate decision was made to make yet another six million dollar expansion for the second season, and it would not be the last investment. Attempting to strengthen the park's anti-modern image, a number of "nostalgia-themed additions" were included over the years (Dollywood, "Yearly"). In 1991, for example, the 1.5 million dollar Eagle Mountain Sanctuary was added, creating an outdoor aviary for non-releasable bald eagles. The next year the Friendship Gardens, a part of the new Showstreet complex, displayed as many as 35,000 flowers and plants in an impressively created gardenscape.

Fig. 13: Dollywood grist mill. Photo by author (2006).

While much emphasis has been placed on strengthening the park's rustic image, over the years park officials have also been highly aware of "greening" Dollywood in order to create the right kind of framework for the displays of antique technologies and handicrafts. As a 2007 press release stated, "when Dollywood was first planned, the natural beauty of the mountain landscape was considered a key attraction of the park. Some areas of Dollywood are still untouched, left just as they have been for centuries, while other areas have been carefully cultivated and developed into a fragrant and beautiful celebration of the land." The press release continued by noting that the two rollercoasters known as the Thunderhead and the Mystery Mine, both situated between two mountains, also had landscape crews paying "special attention to the selection of trees, shrubbery, annuals and ground-covering plants to ensure a setting true to East Tennessee's famous mountain vegetation." And in another area of the park, the "hillsides surrounding the thrilling Tennessee Tornado coaster are landscaped with a very natural look to enhance the ride's theme as an exciting—and beautiful—Smoky Mountains experience that cannot be duplicated. More than 300 native hardwood trees and evergreens, hundreds of shrubs, thousands of grasses and ground covers, and wildflowers have been used to enhance the Tennessee Tornado area's natural, mountainous look" (Dollywood, "Mountains").

Through its gardened and landscaped environments, Dollywood creates an idyllic setting for its displays of Southern Appalachian history and culture. The mix of a supposedly original landscape with manipulated gardenscapes and displays of antique technologies, handicrafts, old cabins, an antique steam train, and grist mills suggests an idealized integration of nature and culture. But in spite of the park's efforts to shape its natural surroundings and attractions into a narrative of the regional past, there is no functioning context for its hyper-rusticity. Rather, history at Dollywood is an attraction located amidst roller coasters, theatres, and souvenir shops. Parton's stated interest in celebrating and honoring the "real people" of the region in which she grew up, for example, is questionable when one considers that until 1995 comedian Billy Baker's performance as Elwood Smooch ranked among the park's most popular shows (Parton 278). Relying on a popular stereotype, Smooch entertained visitors by playing an ignorant and lazy hillbilly wearing an old toilet seat around his neck. Although Baker won much popular praise for his role and eventually got his own show in Pigeon Forge, the act nevertheless perpetuated stereotypes regarding people from the area as being ignorant and backwards. Such portrayals of the Southern Appalachians call into question the extent to which Dollywood is in the business of celebrating or exploiting regional identities, but it may well be a question that has found an answer in co-founder Peter Herschend's recent response to criticism of the company's newly planned Old South theme park on Stone Mountain in Georgia. Confronted with the problematic exclusion of both slavery and the Ku Klux Klan from the park, Herschend offered the reply that his company is in "the entertainment business," not "the business of history" (Sengupta). So much, one is tempted to add, for not leaving history in the hands of Hollywood.

Thus, although Dollywood presents itself as a historically relevant park, it remains an amusement park that draws its popular appeal from a stylized pastoral image of the regional past. History there is an experience, carefully redesigned to appeal to a broad consumer segment by using popular images of a static past. For example, two statues placed on the edge of the 1950s inspired Jukebox Junction area show a girl sitting on a tree stump trying to catch butterflies with her hands, and three children playing in a tree. Frozen in time, the statues, like the rest of the park, convey little sense of any of the hardships or work involved in mountain life. Just as there is nothing that can disturb the playful innocence of the children, there is nothing inside the park that can break the illusion of the regional past that is portrayed at Dollywood.

III

Whereas places like the Tennessee Valley Railroad Museum, the Chattanooga Choo Choo, and Dollywood focus on local, regional, and national audiences, the Jack Daniel's distillery in Lynchburg, Tennessee, caters to a global audience. In 2008, for the first time in the brand's history, Jack Daniel's sold more whiskey abroad than it did on the American market. The company's global success is, at least in part, attributable to the celebrated marketing campaign that, in spite of the company's position as a global multi-million dollar business, has persistently sought to associate the Jack Daniel's brand with the small-town values and ideals of Lynchburg, Tennessee; the town where Jasper Newton "Jack" Daniel was born and where he began distilling the now famous whiskey in the mid-nineteenth century. Through highly stylized, historicized, and localized images of Lynchburg, the Jack Daniel's campaign has successfully sold a distinctly Southern, and, one might argue, rather stereotypical sense of place and history, evoking notions of anti-modernity in order to appeal to a post-war American society increasingly dominated by mass-production.

But the Jack Daniel's campaign has come at a cost to the town of Lynchburg. The use of the local community and its residents in the ads has gradually worked to intertwine the company's marketing strategy with the town's sense of self. The romantic narratives being told over and over again by the ads have permeated the identity of Lynchburg to an extent where reality has been replaced with a kind of permanent communal performance. From the stories being told to visitors on the distillery tours, to the physical space of Lynchburg, all seem to conform to the expectations created by the ads. As the community has sought to live up to the ideals that saturate the advertisements of the county's largest employer and tourist attraction, it has become difficult, if not impossible, to tell apart the *real* Lynchburg from what might be considered the Jack Daniel's experience.

The postcards campaign, as it has become known due to the ads' down-home tone of voice and the fact that they have seemed to address the nation from the small township in Tennessee, was first adopted in 1954 after a popular article in *True* depicted the distillery and Lynchburg as examples of a pre-modernity that was rapidly disappearing elsewhere. Realizing that the article had hit a popular nerve, the company's new ad agency, Gardner Nelson, was quick to adopt the small-town-that-time-had-forgotten narrative as the central metaphor of the new campaign (Holt 10). Featuring local people from Lynchburg instead of expensive models, the campaign consisted of a series of black and white photographs showing old men in overalls sitting on wooden benches playing checkers or participating in other such unhurried activities. One advertisement in the late 1980s, for example, featured a black and white photograph of puppies being sold from the back of a pick-up truck along with the text: "There aren't many towns where coonhound pups are sold on Main Street. But

Lynchburg, Tennessee, is one" ("The Media Business"). The focus on a slower pace, social coherence, and community values correspond to a simplified agrarian South. Setting up an imagined place where residents have supposedly resisted the rampant capitalism and increasing social fragmentation of the 1950s and 1960s, the postcards campaign undoubtedly has appealed to a wide range of American consumers.

The campaign is a deliberate attempt to create a wholesome, country-like image. As noted in one New York Times article, the "familiar magazine ads chronicle the life and times of Lynchburg, Tenn., the bucolic town where Jack Daniel's has been distilled since 1866. With their nostalgic photographs and folksy copy, the ads would surely be sepia-tinted were they not already black-and-white" ("The Media Business"). And the strategy has worked wonders. Market research has shown that the ads appeal to a wide range of American consumers, and in 2004 Jack Daniel's celebrated an unprecedented 50th anniversary of the campaign. As Ted Simmons, CEO of Arnold Worldwide-St. Louis, the advertising agency presently in charge of the campaign, has explained: "Jack Daniel's advertising has always been rooted in the authenticity and timelessness of Lynchburg, its people and how the product is made" ("Jack Daniel's Celebrates").

But just as the Jack Daniel's brand is tied to the imagined values of the small community, Lynchburg's self-image has also become inextricably connected to the advertisements. On the chamber of commerce website, for example, Lynchburg is described as an "old fashioned town, the likes of which you may have thought no longer existed," a place where life in general "moves at a slower pace." Although created by the Moore County Chamber of Commerce, the website features none of the usual boasts about progress or economic opportunity designed to attract new industry to the area. On the contrary, the text sounds like something that might have been copied out of the black and white whiskey ads, suggesting that Lynchburg, on some level, is striving to live up to that idealized small town image.

The brand's anti-modern image is also behind the 2004 venture into the manufacturing of western-style clothes and accessories including belt buckles, cowboy hats, wallets, and watches. Having already tapped into the popular association between whiskey and the Wild West by sponsoring Professional Bull Riders and Professional Rodeo Cowboys Association events, Jack Daniel's has sought to expand its market share beyond the world of the rodeo by adding a clothing line produced by Wrangler ("Jack Daniel's Celebrates"). Although the Wild West image seemingly has little to do with the small town narrative that Jack Daniel's is cultivating, according to Jack Daniel's brand director John Hayes, it was a logical move considering that cowboys "would be more likely to drink Jack Daniel's than the average American," and that whiskey had "always been perceived as a drink with a lot of masculinity to it" ("Jack Daniel's Whiskey"). Whether such claims are true or merely based on popular stereotypes, the cowboy narrative certainly contains the same sort of simple living ideology that runs through most of the brand's advertisements. In that sense, the Jack Daniel's marketing strategy is not unlike the campaign that propelled Marlboro cigarettes to the top of the market via the lonesome, self-contained Marlboro Man.

Professor of marketing Douglas B. Holt sees more than a fleeting connection between the Jack Daniel's campaign and the American frontier mythology. In his study of Jack Daniel's ad campaign as an example of brands as ideological parasites, Holt argues that the Jack Daniel's ads specifically sought to tap into the anti-modernity of the frontier ideology when it shifted to its Lynchburg strategy. Pointing to the gunfighter myth's centrality to American culture, Holt connects the popularity of the postcards campaign with the frontier narratives that permeated American society in the first half of the twentieth century, further arguing that whiskey is a drink that is historically linked to the frontier experience and to representations of that part of American history (Holt 360).

Holt is right in recognizing the link between branding processes and the ideologies that inform them and in pointing out that the pre-modern image has been central to making customers believe that in a time of mass-production and suburban conformity Jack Daniel's is the link to a forgotten past. But the ideological source of the postcards campaign, I would argue, is not the frontier narrative but the perceived rusticity of the late nineteenth and early twentieth century South. Not only does the recent move towards strengthening the brand's image within the consumer segment traditionally attracted to Wrangler and rodeos suggest that the frontier ideology is not at the heart of the brand's original strategy, but whereas the frontier narrative revolves around a social reinvention in a landscape perceived as being void of history and culture, the Jack Daniel's ads, as part of a Southern agrarian narrative, have always emphasized tradition, history, and an overpowering sense of place, which seems logical considering the distillery's location and long-time connection to the town of Lynchburg.

And it is precisely the successful evocation of a sense of place and tradition that has worked wonders around the world. Although there are differences in the extent to which rural Lynchburg is used in the ads—in places like India and China, where many come from poor and rural backgrounds, customers might not find the rustic imagery as appealing as, for example, British consumers—the central message of small-town authenticity is received and understood globally. According to Director of International Marketing Partners Allyson Stewart-Allen, the small-town imagery and the consistent focus on craftsmanship and intimacy may even have meant that the brand has been less likely "to experience boycotting from overseas markets" (Kiley). Thus, when asked about Jack Daniel's, a Chinese consumer explained that he liked "how it's handcrafted and all comes from this one special place" (Kiley).

But the experience of authenticity is not merely cultivated through the ads, it is also a dominant theme in the distillery tour which draws thousands of international visitors to Lynchburg every year. The tour was begun in 1954, the same year as the successful ad campaign, and can, as historian Mark Weiner suggests, be divided into two parts: one that focuses on the "history of the consumer artifact" and another that "explores a past external to the product itself—in this case, the regional past, the nostalgic past of old Tennessee" (231-232). It is hardly surprising that a guided tour is used to further the brand image, but it is interesting that the corporate narrative, formed in unison by advertisements and distillery tour, relies on many of the same signifiers found at places such as the Chattanooga Choo Choo and Dollywood.

Before the tour, tourists gather in the recently built 16,000 square-foot Visitor Center. Built to match the rusticity associated with the Jack Daniel's brand, the center is full of displays showcasing the company's history. Here visitors can gaze at old tools, an antique whiskey delivery truck, and even a checkers board placed on a whiskey barrel with two chairs and an inscription explaining that the game, like distilling whiskey and like life in Lynchburg, is all about patience and allowing things to take their time. As the tour begins, visitors are led into a small theatre where a short seven-minute video presentation introduces the Jack Daniel's brand in the familiar terms, opening with a shot of pastoral surroundings complete with a farm and a corral with some horses. The rest of the tour exposes the visitors to the same message as they walk over the distillery grounds. There they can admire an old fire engine, the handcrafted wooden barrels where the whiskey is stored, the hollow from which the cool water for the whiskey-making process springs, and even Jack Daniel's old office complete with what, supposedly, are the original dusty account books. Emphasizing an adherence to history and traditions, even the guides are a part of the narrative that is being told. Performing as, and conforming to, a stereotypical `good ol' boy´ personae, they embody the easy-going Southern charm, humor, and laid back attitude that is central to the Jack Daniel's story.

And the performance extends beyond the distillery tour and into the streets of Lynchburg. Dependent on the work and tourist dollars generated by Jack Daniel's, the town itself seems to conform to the images of a rural idyll that are propagated by the ads and which guide the expectations of the many visiting tourists. Wherever one ventures in Lynchburg, the Jack Daniel's narrative is visible. Almost every store on the town square sells a mixture of antiques and Jack Daniel's related souvenirs. Old-time rocking chairs and checkers boards, central props of the Jack Daniel's narrative, further the fusion of communal and corporate interests. The Jack Daniel's company has also taken over the regionally famous Miss Mary Bobo's Boarding House, a Southern-style restaurant that opened in 1908 when Miss Mary Bobo took over the historic Salmon Hotel. For a while, as the boarding house was being renovated, the company even hosted the restaurant's guests at its own welcome center. And the impact of the Jack Daniel's advertisements on life in Lynchburg becomes even more apparent when considering the story of Mr. Herb Fanning, a local man who became so famous after appearing in a number of Jack Daniel's ads that he became known as the unofficial mayor of Lynchburg. During the celebration of the postcards campaign's 50th anniversary, Jack Daniel's even put up a statue of Fanning in the town square, a tribute not only to the man himself but also to the interplay between town and company through marketing ("Jack Daniel's Celebrates").

Fig. 14: Jack Daniel's old office is on the distillery tour. Photo by author (2006).

The image of the isolated small Southern community that modernity has passed by is, of course, problematic on several levels. Firstly, it is a narrative that ignores the continued interaction between Lynchburg and the outside world. The Jack Daniel's brand is in fact owned by the Louisville, Kentucky, based Brown-Forman Corporation, one of the nation's largest producers of wine and spirits. Brown-Forman aggressively protects its best-selling brand, and in 1982 sued 13 businesses in Gatlinburg for selling unauthorized reproductions of the distillery's trademark ("Daniel Distillery"). In fact, Jack Daniel's is a far more modern and sophisticated operation than it wants visitors to believe. The actual bottling takes place not in the old warehouse full of wooden barrels that visitors see on the tour, but in an automated bottling facility some miles away (Cone). The narrative also conceals the fact that the heart of the Lynchburg community is in fact a global multimillion dollar industry and a major local employer that, from time to time, has had to make unpopular decisions, as when in 1985 more than 30 workers were laid off as the result of a miscalculation (Weiner 234). But as a plaque at the nearby Jack Daniel's visitors' center notes, there is no reason to spoil a good story with facts.

Secondly, the cultivated rusticity clashes with the reality of modern tourism. Calling the distillery a "theme park wrapped around a production plant," as Edward Cone has done, might exaggerate the point but like Dollywood, both Lynchburg and Jack Daniel's are highly self-conscious about their anti-modern image, and both town and brand rely heavily on a cultural iconography that appeals to urban Americans anxious about the hurriedness of their own lives. The constructed nature of the rustic appearance comes sharply into focus precisely because of the tourism appeal. According to the Jack Daniel's website more than one quarter of a million yearly visitors find their way to the town where the famous whiskey is made. The studiously rustic Visitor Center with rocking chairs lined up on its porches helps to create an ideal atmosphere for relaxing in the shade before and after the tour. But sitting on the porch also creates a heightened awareness of the camera the visitor is carrying, of the tourists flocking in front of the building posing for group photos, and even of the number of high-speed computer stations set up inside of the air-conditioned center that allow visitors to log on to the Jack Daniel's website and register with name, age, and country, supplying the company with valuable marketing information in the process.

Visiting Lynchburg and the distillery on the edge of town is a charming experience, but because the image of Lynchburg and the Jack Daniel's brand are so inextricably linked, and because the distillery plays such an overwhelming role in the formation of the town's identity, there no longer is a clear distinction between the real and the staged. As Weiner argues, apart from being a small town, "Lynchburg also is a tourist attraction, and it would be a mistake to consider the town outside of its interaction with national culture. Just as the traditions of southeast Asian islands have been irreparably altered by the influx of western visitors and anthropologists, so the town of Lynchburg, and the Jack Daniel's distillery, exhibit the imprint of dialectic social dynamics" (Weiner 230). In the Borges fable "On Exactitude in Science" cartographers create a map of the empire the same size as the empire itself. Finding the map useless, however, the following generations discard it, leaving its tattered remains to be "inhabited by Animals and Beggars" (Borges 325).

Through its advertisements the Jack Daniel's brand has created a copy of Lynchburg in an idealized form. But, as opposed to the Borges tale, the copy of Lynchburg has superseded the actual town, leaving its inhabitants to occupy a hyper-reality born of a successful marketing strategy.

Six

Beyond the Hyper-Pastoral: A Conclusion

While in Columbus, Mississippi, researching the Tennessee-Tombigbee Waterway, I had the opportunity to talk with two representatives from the local branch of the Army Corps of Engineers. During the meeting, which took place in the offices of the Tennessee-Tombigbee Waterway Development Authority, I was told an amusing anecdote. The story involved a woman who was staying at the, then newly constructed, DeWayne Hayes Recreation Area on the Tennessee-Tombigbee Waterway just outside of Columbus. Having taken a walk on the campground trails the woman commented to an Army Engineer representative that she was very impressed with the amount of detail they had put into making the area appear natural and that the fake snake she had seen on one of the paved walkways added a wonderful touch of realism.

Of course, the snake—a nonvenomous and rather common Brown Snake—was real and the story was presented as just another example of a tourist ignorant of local conditions, but it stayed with me nonetheless: why would the tourist expect a fake snake in a nature recreation area? And what did her reaction reveal about people's expectations about what nature is and should be? The woman in the story did not find the idea of a fake snake problematic, but immediately accepted it as part of the recreational experience, suggesting that the distinction between the authentic and the inauthentic is becoming increasingly irrelevant. Like the people who visit the hyper-pastoral environments of Dollywood or the Jack Daniel's distillery accepting their pastoral representations as reality, to the tourist at the DeWayne Hayes Recreation Area the presence of artificial wildlife does not automatically threaten the authenticity of her nature experience.

When Leo Marx invented the term middle landscape he did so to account for an adaptation of the traditional pastoral to a rapidly industrializing nation. Faced with such changes the middle landscape incorporated the existence of an external modernity, simultaneously offering an escape from the modern while acknowledging the futility of that very proposition. But tracking the middle landscape only until the publication of F. Scott Fitzgerald's seminal *The Great Gatsby* in 1925, Marx's middle landscape theory does not account for the development of the American pastoral since then. Using Baudrillard's simulacrum theory as a broader theoretical framework, this study has sought to account for that development, by describing how the historic and unintended pastoral, as reality and idea, have served as inspiration for a number of public spaces throughout the modern Tennessee Valley. These places are representative of the pastoral's development from a simulated space to the fragmented reality of the hyper-pastoral.

From the 1890s to the 1930s the pastoral ideal became an integral part of the design of places such as the Biltmore Estate, the Appalachian Trail, and the Great Smoky Mountains National Park. Not merely aesthetically pleasing, the pastoral became a moral landscape central to the reform-hungry progressives, who saw the reintegration of nature and culture as a key to the new society they envisioned. Nostalgic for what the industrial society had ruined, the use of the pastoral was imbued with a utopianism that was to steer modern America in a direction toward egalitarianism and social renewal in nature.

Following in the footsteps of such progressive ideology, in 1933 the Tennessee Valley Authority initiated an economic, social, and cultural transformation of the Tennessee Valley. Part of Roosevelt's New Deal program, the TVA stimulated a region that was lagging behind most of the country economically and technologically. Foremost on the TVA's agenda were economic issues such as flood control, electrification, and navigation, yet under the leadership of the charismatic and visionary Arthur E. Morgan the agency also initiated a host of programs designed to restore the region through the conservation of natural resources and the creation of a tourism-based economy.

Permeating many of the TVA's early programs was a vision of a middle landscape in which nature and technology merged to form a utopian pastoral landscape. The controlled use of TVA-produced fertilizers on exhausted lands, the large-scale reforestation effort that was undertaken to prevent future soil erosion, and creation of test farms to help struggling farmers transform their fields into pastures and orchards, all had the effect of changing the appearance of the landscape. Through educational material and school essay competitions, the TVA further sought to instruct people in the valley about how best to conserve natural resources, and at Norris Dam workers who participated in a forest worker experiment were given small tracts of land in return for taking on forest management responsibilities. In this way participants gained important understanding of how to manage and conserve the forest resources while supplementing their incomes.

The project that probably best sums up the agency's early vision is the small town of Norris. Originally created to provide more permanent housing for the workers on the dam, Norris came to embody many of the principles inherent in the TVA effort. Homes were designed in accordance with the surrounding landscape, using only local materials and designs. Affordable housing sought to eliminate the social barriers usually implied by home ownership; private gardens were eliminated in favor of a larger green public space; citizens were encouraged to undertake private agriculture and instructed in how best to be successful; schoolchildren were educated in gardening and farming; and a central market was established in town allowing locals to trade or buy homegrown products. Furthermore, the town was shielded by a broad greenbelt that prohibited its growth beyond a certain size, and the surrounding highways were kept free of billboards and other privately owned constructions.

Although Norris failed as a social experiment, and by the late 1930s had become a bedroom community for Knoxville-based agency officials, the initiatives begun there were indicative of a pastoral vision that guided many of the TVA's early efforts. Even as late as the 1960s and 1970s—as the agency's stewardship of the Land Between the Lakes recreation area (LBL) shows—the pastoral ideal influenced the agency's decisions in matters of both policy and design.

Often hailed as an experiment in recreation LBL was the result of the damming of both the Tennessee and the Cumberland rivers. The creation of Kentucky Lake by the TVA in 1944 and Lake Barkley by the U.S. Army Corps of Engineers in the late 1950s left only a narrow, largely wooded and sparsely populated peninsula in the area formerly known as the Land Between the Rivers. But as recreational planners soon realized, with 300 miles of shoreline the small strip of land would make a perfect location for a national recreation area, and approximately 67,000 out of the 170,000 acres of land were already in public ownership (Wright 96). In 1961 the TVA sent a proposal regarding the creation of a national recreation area to President Kennedy, and within two years the plan was approved and the agency was authorized to develop the Land Between the Lakes as a demonstration of resource development and environmental education.

In a number of ways LBL was based on the same principles of conservation and recreation as many of the TVA's earlier programs. Natural resources were conserved and carefully managed, while hiking trails, beaches, and boating facilities were constructed to attract the modern camper, including hikers, anglers, bird watchers, nature photographers, and even hunters. Also, in keeping with its educational profile, and in order to attract schools, several large buildings were constructed, including dormitories and kitchens, on a small peninsula in Lake Barkley; a line of nutritious frozen dinners was manufactured to allow for a quick but healthy meal; and during the day teachers and students could roam the nearby woods, conduct classes in an outdoor environment, or go on prearranged trips where they could watch grazing deer or beaver dams at different wildlife stations, plant trees and bushes at forestry stations, and observe farm life at some agricultural stations.

From the beginning, LBL was created with a strong social profile, specifically targeting children from low income groups and inner city environments in nearby cities such as Paducah or Nashville where many parents could not afford to take trips to places such as the Great Smoky Mountains National Park. By offering underprivileged children a chance to experience nature in ways that would otherwise be impossible for many of them, LBL sought to meet what director Frank Smith called one of the greatest needs of the urban poor; namely that of a "clean open space close to home" (Smith 3). Furthermore, in order to avoid a situation where schools that might already be financially burdened would have to bear additional costs that might prohibit them from visiting LBL, free bus rides to and from the area were offered.

Although much was done to create an atmosphere of comfort and luxury at the new campgrounds, LBL's central narrative nevertheless revolved around the traditional pastoral notion of retreating into nature. As an alternative to the modern campgrounds, primitive campgrounds were created for those visitors who wanted a more traditional camping experience. Large parts of the park were even structured to evoke a sense of anti-modernity. The main road running through the recreation area was renamed "The Trace"—a traditional name for many pre-industrial roads and paths—and over time attractions were added to showcase pioneer life in the wilderness. In the late 1970s, for example, a working farm called "The Homeplace" was created, displaying no fewer than 16 log structures on a 40-acre patch of land. The working farm included costumed guides who could tell visitors about farm life in the 1850s as well as a visitor's center with details about the history of the region. And as late as the 1990s an "Elk and Bison Prairie" allowed visitors to experience the area as it had appeared, according to a sign, "during the time of Daniel Boone."

Fig. 15: The Land between the Lakes recreation area—the Homeplace. Photo by author (2006).

The pastoralized space remained popular, but it was changing. The post-war conservatism meant that the visions and the social ideology that had dominated many of the pre-war efforts were disappearing. The pastoral design was still popular, but it was increasingly reproduced outside of its earlier context, and for different reasons. As projects such as the TVA's Tellico Dam and the Army Corps of Engineers' Tennessee-Tombigbee Waterway illustrate, the pastoral was gradually liberated from its traditional meaning as an environmentally balanced and moral space. In both cases, the large-scale public works projects were losing the battle for public interest and with it the battle for funding. As a result, the modern pastoral—in the form of environmental and recreational issues – became a way of appeasing critics and creating a more appealing appearance for projects that might otherwise not have been completed.

To TVA officials in the late 1950s and 1960s, the Tellico Dam came to represent a return to the agency's roots in comprehensive public planning. Having been reduced to a power company in the post-war years, the continued existence of the agency was threatened. On the surface Tellico was similar to earlier projects, and in fact plans for a dam on the Little Tennessee River were first suggested in the mid-1930s. Imitating earlier visions, Tellico was meant to revitalize the region economically and in the process create a new recreational wonderland where locals and tourists alike could spend their leisure time in scenic surroundings.

In reality, however, Tellico lacked the ideological foundation that had guided the agency's efforts during its first years. Whereas the earlier projects had been about conservation and the welfare of the people in the Tennessee Valley, Tellico remained an economic venture designed primarily to supply the TVA with a reason for existing and expanding in post-war America. Also, whereas the TVA had once used recreation and conservation practices to promote a healthy relationship between man and nature, at Tellico environmental concerns often were seen as obstacles and recreational issues functioned primarily to hide the troubling reality of the dam's weak benefit-cost ratio. Furthermore, recreation at Tellico was not a matter of reclaiming eroded and marginal lands for the benefit of the public. The new tourist possibilities came at the price of a beautiful, diverse, and thriving recreational landscape that already attracted thousands of annual hunters, hikers, and anglers.

With the breakdown of the ideologies invested in the pastoralized landscape, and the increasing decontextualization and removal of the pastoral from its original frame of reference, the hyper-pastoral space emerged. In places like the Chattanooga Choo Choo,

Dollywood, and the Jack Daniel's distillery, the pastoral has been reduced to its image value. Where the pastoralized space sought to recreate an imagined environmental and social morality of the pastoral, the hyper-pastoral makes no such pretenses and refers only to its own immediate reality. One such example of the hyper-pastoral is the theme park Dollywood. Revolving around displays of old technologies set within a framework of the eastern Tennessee landscape, the park peddles images of rusticity and regional heritage. It is a narrative not unlike that being told at the Museum of Appalachia, but whereas Irwin's museum is based on a social context, actual experience, and individual people and their stories, Dollywood's use of pastoral metaphors remains glossed-over, decontextualized, and fragmented. A single replica log-cabin, a grist mill, an antique train, and an area of handicraft displays are set within the framework of thundering rollercoaster rides, souvenir shops, and thousands of daily tourists, emphasizing the park's hyperreality. At Dollywood there is no underlying meaning or significance, there is only the pastoral image.

However, many visitors to Dollywood do not seem to experience the park as particularly staged or inauthentic. In fact, comments about the park seem to emphasize quite the opposite: that it feels authentic. Although the pastoral themes of anti-modernity and a retreat into nature are blatantly contradicted by the very origin and formation of Dollywood itself, visitors persist in describing it as less contrived and more relaxed than other theme parks. Not unlike the woman who somehow expected to find an artificial snake at the DeWayne Hayes Recreation Area, the popular response to Dollywood lends credibility to MacCannell's argument that in modern society the genuine experience has been replaced by a staged authenticity, a hyper-pastoralism where the boundary between reality and the reproduced is unclear.

But although the all-encompassing plasticity of the hyper-pastoral may not give the impression that there is anywhere left for the middle landscape to go, its continuing development beyond the image-driven nostalgia of the hyper-pastoral suggests that it remains at the heart of that fundamentally human issue of harmonizing environmental concerns and a modern lifestyle. In response to the perceived threats of the growing environmental crisis, and to the decay of the modern city, new pastoral designs are once again being imbued with a sense of morality in, adapting Terry Gifford's terminology, what might best be described as a post-pastoral space. To Gifford, in literature the post-pastoral genre encompasses a recent move away from the ironic detachment and cynicism of the anti-pastoral descriptions of the landscape. Instead the post-pastoral, Gifford argues, relies on bridging the nature/culture divide by conveying a sense of "nature as culture and of culture as nature" (Gifford 162). In that sense, the post-pastoral might best be described as a rediscovery of the harmony of the pastoral, albeit in new forms, and so seems a useful explanation for the kinds of spaces that might characterize the middle landscape's further development beyond the hyper-pastoral.

Although hardly representative of all six post-pastoral characteristics that Gifford describes, post-pastoral spaces do tend to revolve around a reintegration of human society and nature in new and significant ways. One example of this type of space is the greenway, or greenbelt as it is sometimes known. Greenways are linear, open spaces that often follow rivers and streams, or other such "ecologically significant natural corridors," or even abandoned railways (Little 5). While some greenways are primarily recreational, others revolve around civic, environmental, or economic concerns. In his book *Greenways for America* Charles Little has divided the greenways into five different categories. These include urban riverside greenways that are often part of a reclamation of worn-down river fronts, recreational greenways, ecologically significant corridors that allow for wildlife migration and nature study, scenic and historic routes, and comprehensive greenway systems that incorporate larger areas or several smaller projects in an effort to green regional infrastructures (4-5). In a 1987 report the President's Commission on Americans Outdoors called greenways "our vision of the future" and went on to describe a national system of interconnected "recreation corridors" by which people would be able to travel on foot, bicycle, or even by horse (President's Commission 142).

In spite of its recent popularity, the greenway is not an entirely novel idea. It can, to some extent, be attributed to the works of Frederick Law Olmsted. After having spent a couple of years in California in the early 1860s working, among other things, on the plans for the Yosemite State Park and designs for the Berkeley campus, by 1866 Olmsted returned to the East Coast where, along with his partner Calvert Vaux, he began work on Prospect Park in

Brooklyn. From their work on Central Park, Vaux and Olmsted had noted the inability of a park to refine and civilize the city. Consequently they decided that the construction of the new Brooklyn park should be seen in relation "to the urban environment as a totality" (Schuyler 125). In their preliminary report to the Brooklyn Park Commission in 1866, they proposed the construction of one or more parkways, an idea which they expanded upon in their 1868 report. These parkways would not only make the park available to more people, but would offer "pleasant routes for carriage owners to drive to and from Prospect Park without subjecting themselves to the dangers and disruptions of crowded and poorly maintained city streets" (Davis, "A Pleasant" 231). Furthermore, they might ultimately link a series of green areas from Prospect Park all the way to Central Park in Manhattan across the bay.

To some extent this was an idea that Olmsted had brought with him from his work at Berkeley, where he had planned not only for the inclusion of the Strawberry Creek valley for public parklands, but also for connecting the campus area with the city of Oakland via a pleasure drive through the hills (Little 9). The parkway construction in Brooklyn never did live up to Olmsted and Vaux's original plans, which called for each parkway to be "260 feet wide, with a central roadway, service roads, and pedestrian paths separated by plots of grass and six rows of trees." A system which would "provide for the complete separation of commercial and recreational traffic, create shaded walks and tree-line avenues, increase the desirability of neighboring lots as sites for middle-class houses, and act as a barrier to the spread of fire" (Schuyler 128). But due to a shortage of funding Olmsted and Vaux terminated their involvement by 1873 at a point when much of the park had yet to be constructed and the parkways were still in the planning stages.

By 1880, however, the proposed Ocean and Eastern Parkways had been built—both are today part of the Brooklyn-Queens Greenway that links thirteen urban parks via a number of trails for hiking and biking (Little 11 & 166-172)—and in the first half of the twentieth century the greenway phenomenon was further developed in Benton MacKaye's design for an Appalachian Trail system, as well as in the greenbelt towns of Clarence Stein, Mumford, and Henry Wright. By the 1980s the idea of the greenways began inspiring an effort at the grassroots level when the politics of Reagan as well as the economic recession in the early part of that decade reduced the possibility of purchasing larger tracts of open land for preservation purposes. Forcing conservationists and preservationists to look elsewhere for alternatives to traditional land conservation, the narrow continuous corridors of the greenways suddenly seemed an obvious choice.

Today the greenway movement has taken root on an international scale, manifesting itself in projects all over America, including Chattanooga, Knoxville, Asheville, and Oak Ridge where, in 1992, the city council authorized a citizens group to develop plans for a local greenways concept. The following year the city authorized the plans, and over the course of the last 15 years more than 34 miles of greenways have been constructed in Oak Ridge (Greenways Oak Ridge). In Asheville, North Carolina, the French Broad River became central in the development of the French Broad Greenway, a project that was made possible in part by the environmental reclamation of that river in the wake of the Water Pollution Control Act Amendments of 1972. Born out of a desire to attract tourism, the Asheville Chamber of Commerce, in cooperation with local civic organizations, sought to make the river a central asset to the community by establishing recreational facilities alongside of it (Little 84-90). According to the City of Asheville webpage, local interest groups and politicians are presently in the planning stages of expanding the existing greenways into a system that will weave through much of Asheville (City of Asheville). In Chattanooga, the Riverpark project was developed in order to reduce public costs associated with the urban decay of the riverfront as well as to generate a profit by attracting new business. This was to be accomplished through a redevelopment of the city's riverfront into a green space for public use that would include facilities for fishing, boating, hiking, skating, biking, and picnicking (Little 140-143; Hamilton County). Built between 1982 and 2005, the project was developed by a citizens group and was funded by the city of Chattanooga, Hamilton County, and the Lyndhurst Foundation (Hamilton County). Setting up the private non-profit RiverCity Company to coordinate and financially assist new businesses, already by the late 1980s plans for the construction of hotels and apartment and office buildings had made the greenway a sound investment; and with the construction of the 30 million dollar Tennessee Aquarium, the Riverpark greenway was an astounding success story (Little 140-143).

Another example of post-pastoral design is the green roof. Green roofs, or roof gardens as they are also known, date as far back as the Hanging Gardens in Babylon from 600 B.C. and are also known from vernacular architectures of regions like Scandinavia and Kurdistan (Dunnet and Kingsbury 10). At present, modern and upgraded versions of the green roof are becoming increasingly popular in urban areas all over the world as a way of responding to the challenges posed by climate change. According to the Tennessee plant nursery Growild—a company that has specialized in developing and designing green roofs—the rooftop garden fulfills a wide variety of purposes. It can fulfill a social function as a recreational urban space, while from an environmental point of view the green roof can function as urban wildlife habitat, reduce the heat island effect produced by large asphalted surfaces, help conserve energy by reducing heating and cooling needs, reduce pollution, reduce the noise level, extend the life of the roof, and reduce rainwater runoff and direct it into municipal stormwater drains (Growild). In the case of the more technologically advanced roofs, it is even possible to store and recycle the rainwater for flushing toilets in the building below. Turning the roof of a corporate headquarter or an apartment complex into a "wildlife park in the sky protected from windblown seeds and the vagaries of man," the green roof transforms urban space in much the same way that parks have, albeit on a smaller scale (Brown, "The Window Box"). By utilizing rooftops, the city is inscribed into a narrative of environmental sustainability that has much in common with the utopianism of the pastoralized landscapes.

Roof gardens became popular in the late nineteenth and early twentieth century as the "development of concrete as a roofing material in the mid-1800s" increased the frequency of flat roofs in European and American cities (Dunnet and Kingsbury 9). Flat roofs offered new possibilities for urban recreation by offering potential garden areas to be enjoyed by tenants, workers, or customers. But although the green roof soon became part of the vocabulary of experimental architects like Frank Lloyd Wright and Le Corbusier, early modern roof gardens were constructed mainly for aesthetic and social purposes. This was also the case with one of the era's most famous roof greenings, the five Rockefeller Center roof gardens in

New York City that were installed between 1933 and 1936. The gardens were originally open to the public, but operating at a loss they were eventually restricted for private use (The Greenroof Projects).

The modern green roof movement was launched in the 1960s, by most accounts in 1961 by Reinhard Bornkamm, a researcher at Freie Universität Berlin who began studying the green roofs of the city's older tenement buildings. Inspired by Bornkamm's findings and the development of new waterproofing techniques, by the 1970s green roofs had become popular objects of study in Germany (Grant, *Green Roofs* 13-14). A number of articles and books put emphasis on the desirability of the green roof as a way of improving the urban environment, and in 1975 the Landscape Research, Development and Construction Society (FLL) was established to further study the phenomenon. Two years later the FLL set up a research group to study the green roof, among other things setting standards and guidelines for the construction of green roofs, many of which are still in use even outside of Germany (The Goethe-Institut).

One reason why green roofs have become so hugely popular as urban landscape installations all over the world may have to do with their remarkable adaptability to the modern city. From a twenty-first-century perspective the term *green roof* hardly seems to cover the variety of approaches to the greening of rooftops. To simplify matters somewhat, and since this seems to be how most scholars have approached the topic, I have used the terms *green roof* and *roof garden* interchangeably to cover a broad variety of designs. But according to Christian Werthmann there are important distinctions to be made even between the green roof and the roof garden, or the *extensive* and the *intensive* green roof. Roof gardens, or *intensive* green roofs, he argues, are primarily "installed to be accessed and enjoyed," are more expensive than green roofs, require much maintenance, and are quite heavy due to their deep soil profiles. Contrariwise green roofs, or *extensive* green roofs, often consist of plain grassy areas, weigh little, require a minimum of maintenance, and are often designed to be inaccessible (18). Recently a semi-extensive version of the green roof has been made available which, according to Nigel Dunnet and Noël Kingsbury, is still low weight and requires little care, but which allows for a more diverse variety of plants than the traditional green roof. And then there is the so-called brown roof, which denotes a rooftop covered with substrate or loose materials, and which has purposefully not been planted in order to "recreate typical brownfield conditions" (Dunnet and Kingsbury 4-6). And I have not even mentioned green facades which have some of the same qualities as the green roof.

Green roofs, brown roofs, and the related green facades present a reintroduction of the pastoral as a way of rethinking the urban and the modern. In America current green roofs include growths of "bottlebrush grasses and wild rye atop Chicago City Hall, succulents on the 10-acre roof of Ford's River Rouge truck plant in Dearborn, Mich., flowering chives and dianthus on the Bronx County building in New York, and, at an office building for the Gap in San Bruno, Calif., a coastal oak savannah landscape" (Brown, "The Window Box"). The GAP green roof covers 69,000 square feet at a cost of 60 million dollars, and was designed and built by William McDonough, who is also at the head of the Cradle to Cradle movement that emphasizes new environmentally sensitive approaches to production. In the Tennessee Valley, the TVA has recently included the green roof idea as a basic principle of site development in its "Sustainable Development Guide" for Tennessee Valley businesses.

To Leo Marx, the pastoral embodied the very essence of America, and it did so to the extent that it was remade to fit the American experience in the form of the middle landscape, a peculiarly American and modern version of the pastoral. In the course of the twentieth and twenty-first century the development of the middle landscape from a pastoralized to a hyper-pastoral space, and beyond, testifies to the continuing importance of the pastoral as a moral narrative central to American history and life. Undoubtedly, the middle landscape will continue to develop, simultaneously offering an escape from modernity and reflecting the changes of contemporary society. With the environmental challenges of the twenty-first century in mind, the middle landscape may prove increasingly vital to actualizing a functional integration of nature and culture.

Bibliography

A Guide to the Biltmore Estate. The Biltmore Company, 2004.

Amend, Katherine H. "The Government Builds Houses." *The Nation*. 137. 3570: 644. Print.

Anderson, Larry. *Benton MacKaye: Conservationist, Planner, and Creator of the Appalachian Trail*. Baltimore: The Johns Hopkins University Press, 2002. Print.

Anderson, Nancy K. "'Curious Historical Artistic Data': Art History and Western American Art". In *Discovered Lands Invented Pasts: Transforming Visions of the American West*. Eds. Jules David Prown, et al. New Haven: Yale University Press, 1992. 1-36. Print.

Anderson, Terry H. *The Movement and the Sixties: Protest in America from Greensboro to Wounded Knee*. New York: Oxford University Press, 1995. Print.

Andrews, Richard N. L. *Managing the Environment, Managing Ourselves: A History of American Environmental Policy*. New Haven: Yale University Press, 1999. Print.

"Appeals Court Refuses to Halt Tennessee-Tombigbee Waterway." *The Washington Post* 25 Mar. 1980. Print.

Artman, J. O. "Reforestation." Article sent to *WIRE* and *Teamwork* (Jan. 25 1954). The TVA Forestry Department Files, the U.S. National Archives and Records Administration in Morrow, GA. Print.

"Atlanta, Ga. Newspaper paints bleak picture of Tenn-Tom, Pickens County." *Pickens County Herald* 19 Aug. 1992. Print.

Augé, Marc. *Non-Places: Introduction to an Anthropology of Supermodernity*. Translated by John Howe. Verso, 1995. Print.

Bailey Jr., Tom. "Landowners along Tenn-Tom Oppose Acquisition for Wildlife." *The Commercial Appeal* 8 Apr. 1981. Print.

Baker, Ed. "Flowers in Choo-Choo Gardens Near Peak of Color." *The Chattanooga Times* 6 July 1976. Print

—. "Tennessee Panel Classes Station a Historic Spot." *The Chattanooga Times* 10 Mar. 1973. Print.

Baker, Willis M. "Conservation in the Tennessee Valley." Report by the Forestry Relations Department (no date). The TVA Forestry Department Files, the U.S. National Archives and Records Administration in Morrow, GA. Print.

Ballal, S. K. "Tellico Dam Furor Involves More than Snail Darter." *BioScience*. 27. 9 (Sep. 1977): 586. Print.

Batteau, Allen W. *The Invention of Appalachia*. Tucson: The University of Arizona Press, 1990. Print.

Baudrillard, Jean. *Simulacra and Simulation*. Translated by Sheila Faria Glaser. Ann Arbor: University of Michigan Press, 1994. Print.

Bond, J. Max. "The Educational Program for Negroes in the TVA." *The Journal of Negro Education*. 6. 2 (Apr. 1937): 144-151. Print.

Boorstin, Daniel J. *The Image: A Guide to Pseudo-Events in America*. Vintage Books, 1987. Print.

Borges, Jorge Luis. "On Exactitude in Science." *Collected Fictions*. Trans. Andrew Hurley. Penguin Books, 1998. 325. Print.

Bradley, James G. "The Mystery of Gifford Pinchot and Laura Houghteling." *Pennsylvania History*. 66 (Spring 1999): 199-214. Print.

Brown, Margaret Lynn. *The Wild East: A Biography of the Great Smoky Mountains*. Gainesville: University Press of Florida, 2001. Print.

Brown, Patricia Leigh. "The Window Box Gets Some Tough Competition." *The New York Times* 23 Mar. 2007. Print.

Brumfeld, Carrol. "Engineers Report Challenged as Inadequate on Wildlife Peril." *The Commercial Appeal* 22 June 1972. Print.

Buckles, Eleanor. *Valley of Power*. Creative Age Press, 1945. Print.

Cann, Nancy Morgan. " `Of Friends and Neighbors´: Community Journalism in Norris, Tennessee 1934-1985", thesis presented for the Master of Science Degree, University of Tennessee, Knoxville (December 1985). Print.

CDC – Centers for Disease Control and Prevention. "The History of Malaria, an Ancient Disease." Home Page. Web.

The Chattanooga Choo Choo. "Historic Hotels of America." Home Page. Web.

City of Asheville. "Parks—Greenways." Home Page. Web.
Clapp, Gordon R. *The TVA: An Approach to the Development of a Region*. Chicago: University of Chicago Press, 1955. Print.
Clark, Thomas D. *The Greening of the South: The Recovery of Land and Forest*. Lexington: The University Press of Kentucky, 1984. Print.
Clayton, Cranston. "The TVA and the Race Problem." *Opportunity, Journal of Negro Life*. 12. 4 (Apr. 1934): 111. Print.
Cone, Edward. "Whiskey, Neat." *Baseline* 1 Apr. 2003. Print.
Cooke, Grace MacGowan. *The Power and the Glory: A Novel of Appalachia*. Boston: Northeastern University Press, 2003. Print.
Cowdrey, Albert E. *This Land, This South: An Environmental History*. Lexington: The University Press of Kentucky, 1996. Print.
"Crafts Fair Recalls `Good Old Days'." *The Chattanooga News Free Press* 24 Sep. 1977. Print.
Cutler, Phoebe. *The Public Landscapes of the New Deal*. New Haven: Yale University Press, 1985. Print.
"Daniel Distillery Suing Over Trademark Use." *The Associated Press* 3 Feb. 1982. Print.
"The Darter and the Dam." *The Washington Post* 19 June 1978. Print.
Davidson, Donald. *The Tennessee—The New River: From Civil War to TVA*. J.S. Sanders & Co., 1992. Print.
—. *The Tennessee—The Old River: Frontier to Secession*. J.S. Sanders & Company, 1991. Print.
—. "That This Nation May Endure—The Need for Political Regionalism". *Who Owns America: A New Declaration of Independence*. Eds. Herbert Agar and Allen Tate. ISI Books, 1999. 149-176. Print.
—. "Where Regionalism and Sectionalism Meet." *Social Forces*. 13. 1 (Oct. 1934—May 1935): 23-31. Print.
Davis, Donald Edward. *Where There Are Mountains: An Environmental History of the Southern Appalachians*. Athens: The University of Georgia Press, 2000. Print.
Davis, John P. "A Survey of the Problems of the Negro under the New Deal." *The Journal of Negro Education*. 5. 1 (Jan. 1936): 3-12.
Davis, Timothy. "A Pleasant Illusion of Unspoiled Countryside: The American Parkway and the Problematics of an Institutionalized Vernacular". In *Constructing Image, Identity, and Place: Perspectives in Vernacular Architecture vol. 9*. Eds. Hoagland,

Alison K. & Kenneth A. Breisch. Knoxville: University of Tennessee Press 2003. 228-246. Print.

Devine, Robert S. *Bush Versus the Environment.* New York: Anchor Books, 2004. Print.

"Dollywood, a theme park in Tennessee, is discussed at the International Association of Amusement Parks and Attractions convention in Orlando, Florida." *PR Newswire* 12 Nov. 1986. Print.

Dollywood. "Mountains of Natural Beauty Found In and Around Dollywood." Home Page. Web.

—, "Yearly Expansions." Home Page. Web.

Doster, James F. and David C. Weaver. *Tenn-Tom Country: The Upper Tombigbee Valley.* Tuscaloosa: University of Alabama Press, 1987. Print.

Draper, Earl S. *A Preliminary Survey of Existing and Potential Outdoor Recreation Areas in the Northeast Section of the Tennessee Valley.* Report by the TVA Division of Land Planning and Housing (Apr. 1935). Print.

—. "Authority Plans Useful, Pretty Valley Homes at New Town of Norris, Near Site of Big Dam." *Chattanooga Times* 3 Dec. 1933. Print.

—. "Negro Park Development within the Tennessee Valley Area and on Authority Land in East Tennessee." Letter to General Manager J. B. Blandford, Jr. (Jan. 20 1938). The TVA files, the U.S. National Archives and Records Administration in Morrow, GA. Print.

—. "The TVA Freeway." *American City* Feb. 1934: 20. Print.

Dunn, Durwood. *Cades Cove: The Life and Death of a Southern Appalachian Community 1818-1937.* Knoxville: The University of Tennessee Press, 1988. Print.

Dunnet, Nigel & Noël Kingsbury. *Planting Green Roofs and Living Walls.* Portland: Timber Press, 2004. Print.

Eco, Umberto. *Travels in Hyperreality.* Translated by William Weaver. New York: Harcourt Brace Jovanovich, 1986. Print.

Elliot, Reed A. Untitled memorandum to L. J. Van Mol attached to "The Tellico Project on the Little Tennessee River." Report by the Project Planning Branch (Oct. 18 1963). The TVA Engineering Report Files, the U.S. National Archives and Records Administration in Morrow, GA. Print.

—. "The Tellico Project of the Tennessee Valley Authority." Report by the Information Office (May, 1965). The TVA Engineering Report Files, the U.S. National Archives and Records Administration in Morrow, GA. Print.

—. "The Tellico Project on the Little Tennessee River." Report by the Project Planning Branch (Oct. 1963). The TVA Engineering Report Files, the U.S. National Archives and Records Administration in Morrow, GA. Print.

Engelhardt, Elizabeth S. D. *Tangled Roots of Feminism, Environmentalism, and Appalachian Literature*. Athens: Ohio University Press, 2003. Print.

"The Engineering of Deceit." *The New York Times* 4 Dec. 1978. Print.

Eschemeyer, R. W. "Fishing on the `Great Lakes´ of the South." Draft for *Chattanooga Times* article (Aug. 1940). The TVA Forestry Department Files, the U.S. National Archives and Records Administration in Morrow, GA. Print.

"Farmer Vanderbilt." *Asheville News and Hotel Reporter*. 20 Feb. 1897: 1. Print.

Farrell, James J. *One Nation Under Goods: Malls and the Seductions of American Shopping*. Washington: Smithsonian Books, 2003. Print.

Featherstonhaugh, George William. *A canoe voyage up the Minnay Sotor; with an account of the lead and copper deposits in Wisconsin; of the gold region in the Cherokee country; and sketches of popular manners; &c. &c. &c.* London: R. Bentley, 1847. *Library of Congress*. Web.

Fields, Norma. "$4 Million Price Tag on 4 Tenn-Tom Recreational Projects." *Daily Journal* 25 June 1975. Print.

"500 Get Jobs at Townsite." *Knoxville Journal* 16 Dec. 1933. Print.

Foresta, Ronald. "Transformation of the Appalachian Trail". *The Geographical Review*. 77. 1 (Jan. 1987): 76-85. Print.

Frank, Bernard. "Notes on Norris Reservoir Program of TVA Forestry Division." Report by the Land Planning and Housing Division (Mar. 16 1936). The TVA Forestry Department Files, the U.S. National Archives and Records Administration in Morrow, GA. Print.

"Getting Up Steam." *The Chattanooga Times* 26 Feb. 1973. Print.

Gibson, Paul. "Boondoggle."*Forbes Magazine* 1 Mar. 1976: 55-56. Print.

Gifford, Terry. *Pastoral*. London: Routledge, 1999. Print.
The Goethe-Institut. "Green Roofs a German Export Hit: from the nation of thinkers and poets to the rooftops of the world." Home Page. Web.
Govan, Gilbert E. "The Chattanooga Union Station." *The Tennessee Historical Quarterly*. 29. 4 (Winter 1970-1971): 372-378. Print.
Grant, Gary. *Green Roofs and Facades*. IHS BRE Press, 2006. Print.
Grant, Nancy L. *TVA and Black Americans: Planning for the Status Quo*. Philadelphia: Temple University Press, 1990. Print.
Gray, Aelred J. "Norris, Tennessee: A History of America's First Self-Contained Greenbelt Town." Unpublished abstract (July 1986). The Aelred "Flash" Gray Papers in the University of Tennessee Special Collections Library, Knoxville, TN. Print.
Greene, Johnny. "Selling off the Old South." *Harpers* Apr. 1977: 39-58. Print.
Greiner, Patricia. "Radical Environmentalism in Recent Literature Concerning the American West." *Rendezvous: Journal of Arts and Letters*. 19. 1 (Fall 1983):8-15. Print.
The Greenroof Projects Database. "Rockefeller Center Roof Gardens." Home Page. Web.
Greenways Oak Ridge. Home Page. Web.
Growild. "Green Roofs." Home Page. Web.
Hamilton County. "TN Riverpark—History." Home Page. Web.
Hargrove, Erwin C. *Prisoners of Myth: The Leadership of the Tennessee Valley Authority 1933-1990*. Princeton, NJ: Princeton University Press, 1994. Print.
Herndon, G. Melvin. "A 1796 Proposal for a Tennessee-Tombigbee Waterway." *Alabama Historical Quarterly*. 37. 3 (1975): 176-182. Print.
Hershey, John W. "Forest Tree Crops and Their Part in the Tennessee Valley." Report by the Department of Forestry Relations (Nov. 1934). The TVA Forestry Department Files, the U.S. National Archives and Records Administration in Morrow, GA. Print.
"The Holdup of No. 4501." *Ties, Southern Railway System Magazine* June 1964:16-17. Print.
Holt, Douglas B. "Jack Daniel's America: Iconic brands as ideological parasites and proselytizers." *Journal of Consumer Culture*. 6. 3 (2006): 355-377. Print.

"Homes of Workers In TVA to Follow Natural Designs." *Christian Science Monitor* 25 Nov. 1933. Print.

Howes, Robert H. "Opportunities for the Development of Negro Recreation Facilities on Tennessee Valley Authority Property." Report by the Department of Regional Planning Studies (July 19, 1938). The TVA files, the U.S. National Archives and Records Administration in Morrow, GA. Print.

Huffman, Alan and Michael Rejebian. "Tenn-Tom: Waiting for its ship to come in." *Clarion-Ledger* 26 Feb. 1989. Print.

"Huge recreation area possible on Tenn-Tom." *Amory News Advertiser* 20 Nov. 1977. Print.

Irwin, John Rice. "The Museum of Appalachia: A Living Mountain Village—Norris, Tennessee." Brochure published by the Museum of Appalachia. Print.

—. *The Museum of Appalachia Story*. Schiffer Publishing, 1987. Print.

"Jack Daniel's Celebrates 50 Years of Legendary Advertising; Celebration Includes Statue Unveiling in Lynchburg, Tennessee." *Business Wire* 9 Sep. 2004. Print.

Jack Daniel's. "Exploring Lynchburg." Home Page. Web.

"Jack Daniel's Whiskey now making Western clothes to boost brand." *The Associated Press State & Local Wire* 15 Jan. 2004. Print.

Jones, Finn-Olaf. "The Danish Soul of That Town in *Sideways*." *The New York Times* 4 June 2006. Print.

Jones, Madison. *A Buried Land*. Second Chance Press, 1987. Print.

Jolley, Harley E. *The CCC in the Smokies*. Pamphlet published by the Great Smoky Mountains Association, 2001.

Kephart, Horace. *Our Southern Highlanders: A Narrative of Adventure in the Southern Appalachians and a study of Life Among the Mountaineers*. Land of the Sky Books, 2001. Print.

Kiley, David. "Jack Daniel's International Appeal". *BusinessWeek* 10 Oct. 2007. Print.

King, Seth S. "Waterway Barely Survives in House." *The New York Times* 24 July 1981. Print.

Kline, Ronald R. *Consumers in the Country: Technology and Social Change in Rural America*. Baltimore: The Johns Hopkins University Press, 2000. Print.

Kovacs, Peter and Brett Guge. "The Tenn-Tom Waterway: Ten Years of Controversy." *Birmingham News* 27 Nov. 1978. Print.

"Landowners jeer proposal to buy land for Tenn-Tom." *Jackson Daily News* 12 Apr. 1981. Print.

"Last Tellico Residents Leave." *The Washington Post* 14 Nov. 1979. Print.

Leopold, Aldo. *A Sand County Almanac and Sketches Here and There*. Oxford: Oxford University Press, 1989. Print.

Lilienthal, David E. *The Journals of David E. Lilienthal: Volume 1: The TVA Years 1939-1945*. Harper & Row Publishers, 1964. Print.

—. *TVA—Democracy on the March*. New York: Harper & Brothers Publishers, 1944. Print.

Linde-Laursen, Anders. "'Unsettled' Solvang, Danish Capital of America: A Photoessay." *South Atlantic Quarterly*. 98. 4 (1999): 782-799. Print.

Little, Charles E. *Greenways for America*. Baltimore: The Johns Hopkins University Press, 1995.

Lucarelli, Marc. "Benton MacKaye's Appalachian Trail: Imagining and Engineering a Landscape". In: *Technologies of Landscape: From Reaping to Recycling*. Ed. David E. Nye. Boston: University of Massachusetts Press, 1999. 207-217. Print.

Lynn, Patrick. "Facility is Unveiled by Engineers." *Commercial Dispatch* 12 Sep. 1976. Print.

MacCannell, Dean. *The Tourist: A New Theory of the Leisure Class*. Schocken Books, 1989. Print.

MacKaye, Benton. "The Appalachian Trail: A Guide to the Study of Nature." *The Scientific Monthly*. 34. 4 (Apr. 1932): 330-342. Print.

—. "An Appalachian Trail: A Project in Regional Planning." Republished in Anderson, Larry. *Benton MacKaye: Conservationist, Planner, and Creator of the Appalachian Trail*. Baltimore: The Johns Hopkins University Press, 2002. 371-379. Print.

—. "Foreword." In: Fisher, Ronald M. *The Appalachian Trail*. National Geographic Society, 1972. Print.

—. *The New Exploration: A Philosophy of Regional Planning*. Urbana, Illinois: The University of Illinois Press, 1990. Print.

Martin, Christopher Brenden. "Selling the Southern Highlands: Tourism and Community Development in the Mountain South." Diss. University of Tennessee, Knoxville, 1997. Microform.

Marx, Leo. *The Machine in the Garden: Technology and the Pastoral Ideal in America.* New York: Oxford University Press, 1970. Print.

Massey, David. "Over a Barrel: Southern Waterways and the Army Corps of Engineers." *Southern Exposure.* 8. 1 (1980): 92-100. Print.

McDonald, Emily. "Dollywood will offer craft classes." *The Chattanooga Times*, 12 Jan. 1996. Print.

McDonald, Michael J. & John Muldowny. *TVA and the Dispossessed: The Resettlement of Population in the Norris Dam Area.* Knoxville: University of Tennessee Press, 1982. Print.

"The Media Business: Advertising; Jack Daniel's Strategy is Turning 35." *The New York Times* 20 July 1989. Print.

Mencken, H.L. "The Sahara of the Bozart." *New York Evening Mail* 13 Nov. 1917. Print.

Metropolitan Lynchburg / Moore County Chamber of Commerce. "Home." Home Page. Web.

Miller, Char. "Char Miller on Gifford Pinchot, Photographer." *Environmental History*. 8. 2 (April 2003): 312-315. *The History Cooperative*. Web.

—. *Gifford Pinchot and the Making of Modern Environmentalism.* Washington, D.C.: Island Press, 2004. Print.

Mitchell, Carolyn. "Septemberfest in Garden at Choo-Choo." *The Chattanooga Times* 24 Sep. 1977. Print.

Morgan, Arthur E. *The Making of the TVA*. Buffalo, NY: Prometheus Books, 1974. Print.

Morris, Philip. "Don't Just Save the Past, Use It." *Southern Living* Feb. 1974: 69-85. Print.

Moser, Don. "Dig they must, the Army Engineers, securing allies and acquiring enemies." *Smithsonian*. 7. 9 (Dec. 1976): 40-51. Print.

Mould, David. "Landowners Bristle At Tenn-Tom Wildlife Plan." *The Commercial Appeal* 11 Apr. 1981. Print.

Muir, John. *A Thousand Mile Walk to the Gulf.* In: *The Wilderness Journeys*. Canongate Classics, 1999. Print.

Murfree, Mary Noailles. "Drifting Down Lost Creek." In: *In the Tennessee Mountains*. The Gregg Press, 1968. Print.

—. *His Vanished Star*. Houghton, Mifflin and Company, 1894. Print.

"The Museum of Appalachia named an official affiliate of the Smithsonian Institution." *Associated Press Worldstream* 25 May 2007. Print.

Neely, Jack. "Tellico Dam Revisited." *Metro Pulse*. 14. 50 (9 Dec. 2004). Print.

"New Government Sponsored Community Will Unite Farm and Industrial Life." *Tennessean* 25 Nov. 1933. Print.

"New Station is Complete." *The Chattanooga Times* 10 Dec. 1908. Print.

Nye, David E. *American Technological Sublime*. Cambridge, MA: MIT Press, 1994. Print.

—. *Electrifying America: Social Meanings of a New Technology*. Cambridge, MA: MIT Press, 1997. Print.

"Odyssey of the 4501." *Ties, Southern Railway System Magazine* Sep. 1966: 11-13. Print.

Olmsted, Frederick Law. "Yosemite and the Mariposa Grove: A Preliminary Report, 1865." Yosemite Online. Home Page. Web.

Page, Irene. "Tenn-Tom's anniversary passes unheralded." *The Northeast Mississippi Daily Journal* 14 Jan. 1987. Print.

Parton, Dolly. *Dolly: My Life and Other Unfinished Business*. New York: Harper Collins, 1994. Print.

"Passing From Local Scene." *The Chattanooga News Free Press* 16 Aug. 1970. Print.

Pedersen, Kristian. "På besøg i Dolly Partons baghave." *Berlingske Tidende* 9 Mar. 2007. Print.

Perry, George S. "Revised Forest Development Section Policy Statement". Report by George S. Perry, Chief of Forest Development Section (8 Oct. 1936). The TVA Forestry Department Files, the U.S. National Archives and Records Administration in Morrow, GA. Print.

Peterson, Cass. "The Fizzling of 200-Year-Old Dream; As Shortcut to Gulf, Tenn-Tom Waterway Failing to Bring Prosperity." *The Washington Post* 26 Dec. 1986. Print.

Pierce, Dan. "Anne M. Davis." *The Tennessee Encyclopedia of History and Culture*. Web.

Pierce, Daniel S. *The Great Smokies: From Natural Habitat to National Park*. Knoxville: University of Tennessee Press, 2001. Print.

Pinchot, Gifford. *Biltmore Forest: An Account of Its Treatment, and the Results of the First Year's Work.* R.R. Donnelley & Sons, Co., 1893. Print.

—. *Breaking New Ground.* Seattle: University of Washington Press, 1972. Print.

—. *The Conservation Diaries of Gifford Pinchot.* Ed. Harold K. Steen. The Forest History Society, 2001. Print.

President's Commission on Americans Outdoors. *Americans Outdoors: The Legacy, the Challenge.* Island Press, 1987. Print.

"Railroad Unit Turns Ground." *The Chattanooga News Free Press* 5 Jan. 1962. Print.

"Recreation Overshadows Commercial Use on Waterway." *Starkville Daily News* 12 July 1985. Print.

Reisner, Marc. *Cadillac Desert: The American West and its Disappearing Water.* London: Pimlico, 2001. Print.

Relph, Edward. *Place and Placelessness.* Pion Limited, 1983. Print.

"Report of National Resources Committee". Report included in "Memorandum from the President" (Washington: 24 Apr. 1939). The records of the Office of the Chief of Engineers, the U.S. National Archives and Records Administration in Morrow, GA. Print.

"Report of the Committee Composed of Representatives of the Tennessee Valley Authority, Tennessee State Department of Conservation, and Tennessee State Planning Commission Concerning the Proposal for Creating a Negro Park on TVA lands in East Tennessee." Report (6 Dec. 1937). The TVA files, the U.S. National Archives and Records Administration in Morrow, GA. Print.

"Restaurant at Terminal." *The Chattanooga News Free Press* 12 Feb. 1972. Print.

Reuter, Lisa. "Museum of Appalachia a Vivid Walk Into Pioneer Past." *Columbus Dispatch* 6 March 1994. Print.

Richards, Edward C.M. "A Forest Policy for the Tennessee Valley Authority." Report by E.C.M. Richards (8 Mar. 1934). The TVA Forestry Department Files, the U.S. National Archives and Records Administration in Morrow, GA. Print.

—. "Projects over the Valley as a Whole." Memo from E.C.M Richards to William L. Killebrew (9 Feb. 1937). The TVA

Forestry Department Files, the U.S. National Archives and Records Administration, Morrow, GA. Print.

Rome, Adam. *The Bulldozer in the Countryside: Suburban Sprawl and the Rise of American Environmentalism.* New York: Cambridge University Press, 2001. Print.

Roosevelt, Franklin D. *The Public Papers and Addresses of Franklin D. Roosevelt: Volume 2—The Year of Crisis, 1933.* New York: Russel & Russel, 1938. Print.

Roper, Laura Wood. *FLO: A Biography of Frederick Law Olmsted.* Baltimore: The Johns Hopkins University Press, 1973. Print.

Rowe, Peter G. *Making a Middle Landscape*. Cambridge, MA: MIT Press, 1991. Print.

Runte, Alfred. *National Parks: The American Experience*. Lincoln: University of Nebraska Press, 1992. Print.

Rybczynski, Witold. *A Clearing in the Distance: Frederick Law Olmsted and America in the Nineteenth Century*. New York: Scribner, 1999. Print.

Sanford Charles L. "Untitled Review." *American Quarterly*. 17. 2 (Summer 1965): 272-276. Print.

Sayre, Maggie Lee. *`Deaf Maggie Lee Sayre′: Photographs of a River Life*. Ed. Tom Rankin. Oxford: University Press of Mississippi, 1995. Print.

Schaffer, Daniel. "Environment and TVA: Toward a Regional Plan for the Tennessee Valley, 1930s." *Tennessee Historical Quarterly*. 43. 4 (1984): 333-354. Print.

—. "Norris: Historic Town And How It Got The Way It Is."In: *Norris, TN 1933-1983: Norris Kaleidoscope: A Not-Quite-Complete Report Of Events, Activities, People In A Not-Quite-Typical Tennessee Town During Its First Half Century*. Published by the *Courier-News*, 1983. Print.

Scheider, Fred. "Steam Choo-Choo Arrives Despite `Holdup,′ Hotbox." *The Chattanooga Times* 7 June 1964. Print.

Schenck, Carl Alwin. *Cradle of Forestry in America: The Biltmore Forest School, 1898-1913*. The Forest History Society, 1998. Print.

Schmidt, William B. "Tombigbee Waterway Use Falls Short of Forecasts." *The New York Times* 16 Feb. 1986. Print.

Schuyler, David. *The New Urban Landscape: The Redefinition of City Form in Nineteenth Century America*. Baltimore: The Johns Hopkins University Press, 1988. Print.

Sears, John. *Sacred Places: American Tourist Attractions in the Nineteenth Century*. Oxford: Oxford University Press, 1989. Print.

Segal Howard P. "Leo Marx's Middle Landscape: A Critique, a Revision, and an Appreciation." *Reviews in American History*. 5. 1 (Mar. 1977): 137-150. Print.

Selznick, Philip. *TVA and the Grass Roots: A Study of Politics and Organization*. Berkeley: University of California Press, 1984. Print.

Sengupta, Somini. "Georgia Park Is to Hail `Southern Spirit´." *The New York Times* 8 Oct. 2000. Print.

Shearer, John. "N.Y. Architect Barber Created Terminal Station Landmark." *The Chattanooga News Free Press* 22 Jan. 1945. Print.

Sinclair, Ward. "Environmentalists Warn Carter of Backlash Over Tellico Dam." *The Washington Post* 27 Sep. 1979. Print.

—. "Spurious Analysis Aids Water Project." *The Washington Post* 5 February 1979. Print.

Smith, Frank E. *Land Between the Lakes: Experiment in Recreation*. Lexington: The University Press of Kentucky, 1971. Print.

Smyth, Mitchell. "Hello Dollywood." *The Toronto Star* 12 May 1990. Print.

Smyth, Mitchell. "Hillbilly Heaven." *The Toronto Star* 25 July 1992. Print.

Southern Highland Craft Guild. "History of the Guild." Home Page. Web.

The Southwest Virginia Agricultural Association. *Building a New Dominion*. The Southwest Virginia Agricultural Association (Jan. 1943). Print.

Spence, Mark David. *Dispossessing the Wilderness: Indian Removal and the Making of National Parks*. Oxford: Oxford University Press, 1999. Print.

Spirn, Anne Whiston. "Constructing Nature: The Legacy of Frederick Law Olmsted." In: *Uncommon Ground: Rethinking the Human Place in Nature*. Ed. William Cronon. New York: W.W. Norton & Company, 1996. 91-113. Print.

Starnes, Richard D. *Creating the Land of the Sky: Tourism and Society in Western North Carolina*. Tuscaloosa: The University of Alabama Press, 2005. Print.

"Station Project Praised By City, County Officials." *The Chattanooga Times* 30 Apr. 1972. Print.

"'Steam Train to Oz′ to Include Exhibits as Well as Rides Oct. 7." *The Chattanooga Times* 24 Apr. 1961. Print.

Stine, Jeffrey K. "Environmental Politics in the American South: The Fight over the Tennessee-Tombigbee Waterway." *Environmental History Review*. 15. 1 (Spring 1991): 1-24. Print.

—. *Mixing the Waters: Environment, Politics, and the Building of the Tennessee-Tombigbee Waterway*. Akron, OH: University of Akron Press, 1993. Print.

—. "The Tennessee-Tombigbee Waterway and the Evolution of Cultural Resources Management". *The Public Historian*. 14. 2 (Spring 1992): 7-30. Print.

Stokes, Alison. "Welcome to Dollywood." *South Wales Echo* 4 October 2008. Print.

Sutter, Paul. *Driven Wild: How the Fight against Automobiles Launched the Modern Wilderness Movement*. Seattle: University of Washington Press, 2002. Print.

Talbert, Roy, Jr. *FDR's Utopian: Arthur Moran of the TVA*. Jackson: University of Mississippi Press, 1987. Print.

"Tellico Dam Work Halted." *The Washington Post* 25 February 1977. Print.

"Tellico Follies." *The Washington Post* 27 Sep. 1979. Print.

Tennessee Valley Authority. "Mr TVA." Home Page. Web.

—. *Non-Urban Outdoor Recreation: An Analysis of its Functions, Forms, and Types of Areas*. Report by the Division of Land Planning and Housing, the Recreation and Conservation Section (July 1935). Print.

—. *Recreational Development of the Southern Highlands Region: A Study in the Use and Control of Scenic and Recreational Resources*. Report by the Department of Regional Planning Studies (July 1938). Print.

—. "Review of Report and Comments on a Negro Park Development on Norris Lake." Report by the Department of Regional Planning Studies (20 Jan. 1938). The TVA files, the U.S. National Archives and Records Administration in Morrow, GA. Print.

—. "Site Development." *Sustainable Development Guide*. Home Page. Web.

—. "TVA—CCC Camps in Valley-Wide Attack on Erosion." Unspecified document (April 1936). The TVA Forestry Department Files, the U.S. National Archives and Records Administration in Morrow, GA. Print.

—. "Waste Lands to Woodlands". Typescript, film manuscript (Nov. 1941). The TVA Forestry Department Files, the U.S. National Archives and Records Administration in Morrow, GA. Print.

Tennessee Valley Railroad Museum. "The `Copper Basin´ Special." Brochure published by the Tennessee Valley Railroad Museum. Print.

Tennessee-Tombigbee Waterway Development Authority. "The Tennessee-Tombigbee Story." Promotional material published by the Tennessee Tombigbee Waterway Development Authority, Jan. 1969. Print.

—. "The Tennessee-Tombigbee Waterway." Home Page. Web.

—. "The Tennessee-Tombigbee Waterway: Twenty Years of Economic Progress—And Still Counting." Brochure published by the Tennessee-Tombigbee Waterway Development Authority. Print.

—. "Tennessee-Tombigbee Waterway—Your Waterway." Promotional material published by the Tennessee-Tombigbee Waterway Development Authority. Print.

—. "Touring the Tenn-Tom Waterway." Brochure published by the Tennessee-Tombigbee Waterway Development Authority. Print.

"Tenn-Tom: A Recreational Paradise." *Daily Corinthian* 5 Dec. 1977. Print.

"Terminal Station Was Constructed in 1909 in Elegance of Famed New York City Bank." *The Chattanooga Times* 30 Apr. 1972. Print.

Thurmond, A.K. "Material for 'American Conservation in Picture and Story'." Statement by A.K. Thurmond, Forestry Relations Department (1 July 1941). The TVA Forestry Department Files, the U.S. National Archives and Records Administration in Morrow, GA. Print.

Turner, Paul Venable. *Campus: An American Planning Tradition*. Cambridge, MA: MIT Press, 1984. Print.

"TVA's Bitter Victory." *Newsweek* 26 Nov. 1979. Print.

Uhlenbrock, Tom. "Dolly Parton's theme park is a tribute to her Tenn. Roots." *The Philadelphia Inquirer* 17 Sep. 2006. Print.

US Army Corps of Engineers. "DeWayne Hayes Recreation Area." Brochure published by the US Army Corps of Engineers, Mobile District (revised edition, 2003). Print.

—. "Piney Grove Campground." Brochure published by the US Army Corps of Engineers, Mobile District. Print.

—. "Town Creek Campground." Brochure published by the US Army Corps of Engineers, Mobile District (revised edition, 2003). Print.

War Department. "Survey report: Waterway Connecting the Tombigbee and Tennessee Rivers." Report by the War Department, United States Engineer Office (1939). The records of the Office of the Chief of Engineers, the U.S. National Archives and Records Administration in Morrow, GA. Print.

Warren, Robert Penn. *Flood: A Romance of Our Time*. New York: Random House, 1963. Print.

"Waterway Project is Seen Endangered by Viet War." *Jackson Daily News* 23 Jan. 1968. Print.

Watkins, T. H.. *The Great Depression: America in the 1930s*. Boston: Back Bay Books, 1993. Print.

Weiler, Joseph and Shirley Downing. "A New Lifestyle in the Red Clay Hills." *The Commercial Appeal* 21 April 1974. Print.

Weiner, Mark. "We Are What We Eat; or, Democracy, Community, and the Politics of Corporate Food Displays." *American Quarterly*. 46. 2 (June 1994): 227-250. Print.

Werthmann, Christian. *Green Roof—A Case Study: Michael van Valkenburgh Associates' Design for the Headquarters of the American Society of Landscape Architects.* New York: Princeton Architectural Press, 2007. Print.

Wheeler, William Bruce & Michael J. McDonald. *TVA and the Tellico Dam, 1936-1979: A Bureaucratic Crisis in Post-Industrial America*. Knoxville: University of Tennessee Press, 1986. Print.

Wild River. Dir. Elia Kazan. Perf. Montgomery Clift and Lee Remick. Twentieth Century-Fox Film Corporation, 1960. Film.

Wolfe, Thomas. *The Hills Beyond*. New York: Harper & Brothers Publishers, 1941. Print.

Wright, Ann W. et al.. "The Tennessee Valley Authority's Management of Land Between the Lakes: An Overview". In *Land*

Between the Lakes, Kentucky and Tennessee: Four Decades of Tennessee Valley Stewardship. Eds. Edward W. Chester & James S. Fralish. Clarksville, TN: Austin Peay State University, 2002. 95-106. Print.

Index

Lightning Source UK Ltd.
Milton Keynes UK
09 August 2010

158119UK00003B/129/P

9 789042 030572